FAITH AND FOSSILS

FAITH AND FOSSILS

The Bible, Creation, and Evolution

Lester L. Grabbe

William B. Eerdmans Publishing Company
Grand Rapids, Michigan

Wm. B. Eerdmans Publishing Co.
2140 Oak Industrial Drive N.E., Grand Rapids, Michigan 49505
www.eerdmans.com

Published 2018
Printed in the United States of America

ISBN 978-0-8028-6910-4

Library of Congress Cataloging-in-Publication Data

Names: Grabbe, Lester L., author.
Title: Faith and fossils : the Bible, creation, and evolution / Lester L. Grabbe.
Description: Grand Rapids : Eerdmans Publishing Co., 2018. |
Includes bibliographical references and index.
Identifiers: LCCN 2017050292 | ISBN 9780802869104 (pbk. : alk. paper)
Subjects: LCSH: Creation – Biblical teaching. | Evolution (Biology) – Religious aspects – Christianity – Biblical teaching. | Theological anthropology (Christianity) – Biblical teaching.
Classification: LCC BS651 .G764 2018 | DDC 231.7/652 – dc23
LC record available at https://lccn.loc.gov/2017050292

To
Irwin R. Spear (1924–2002)

and
Simon Conway Morris

and
All My Friends on Different Sides of These Questions

CONTENTS

PART II
EVANGELICALS AND EVOLUTION

PART III
ADAM AND HUMAN ANCESTRY

PREFACE

My aim in this book is to discuss the question of creation and evolution from a biblical point of view, as a scholar who specializes in the Bible. There are many books on the question from a scientific point of view. Most scientists write from their area of expertise (though this is by no means always the case); few possess specialist knowledge of the Bible. When I talk about the Bible, I sometimes cite secondary literature, but most of the time I am working with primary sources: the Hebrew, Aramaic, and Greek texts of the Hebrew Bible and the New Testament, the Dead Sea Scrolls, the relevant early translations (the Greek translations of the Hebrew Bible [the Septuagint], and the targumim), and the range of early Jewish literature in a variety of languages. When a source in another modern or ancient language is quoted, the translation is mine unless otherwise indicated.

When I discuss the Bible, I do so as a specialist. Here and there, however, it will be necessary to bring up scientific material. In such cases, I have done my best to research the latest information and views, including consultation of books written by specialists, academic journals, and the opinions of experts. But I recognize that here I am not a specialist and can only pass on information from specialist sources. Nevertheless, I have expended a lot of time and effort to obtain up-to-date information and have also attempted to dig out the views of scientists who have a religious belief. The sources are indicated in the notes. I know that there will inevitably be some errors in my discussion; I only hope and trust that they are not fundamental to the argument. Professor Conway Morris kindly looked over some of the material and saved me from some mistakes (most of them small, happily), but he is not responsible for any that remain. Otherwise, I present the results of my research in the expectation that any mistaken data will not invalidate the argument presented. No doubt expert reviewers will tell me if that is not the case.

Finally, I wish to thank Trevor W. Thompson, acquisitions and development editor, and Michael N. Thomson, senior acquisitions editor at Eerdmans. Together we have worked hard to produce a book that will reach a wide audience, yet without compromising the integrity of the arguments or slighting the data needed to support those arguments. I can only hope that we have succeeded.

LESTER L. GRABBE
Kingston-upon-Hull, England

ILLUSTRATIONS AND TABLES

Illustrations

Tables

ABBREVIATIONS

ABS	Archaeology and Biblical Studies
AGJU	Arbeiten zur Geschichte des antiken Judentums und des Urchristentums
AOAT	Alter Orient und Altes Testament
AS	Assyriological Studies
BAR	*Biblical Archaeology Review*
BCE	Before the Common Era
BJS	Brown Judaic Studies
CE	Common Era
CHANE	Culture and History of the Ancient Near East
DJD	Discoveries in the Judaean Desert
JSOTSup	Journal for the Study of the Old Testament Supplement Series
LXX	Septuagint
MC	Mesopotamian Civilizations
MT	Masoretic Text
SBLDS	Society of Biblical Literature Dissertation Series
SBLSBS	Society of Biblical Literature Sources for Biblical Study
SP	Samaritan Pentateuch
TSAJ	Texte und Studien zum antiken Judentum
VTSup	Supplements to Vetus Testamentum

FAITH AND FOSSILS

PART I

A Scholar's Story

1

FROM BIBLE BELT TO BIBLE SCHOLAR

The Journey Begins

I am a biblical scholar. I have devoted my life to the study of the Bible: literature, languages, and history. My interests span from ancient Near Eastern peoples (the Sumerians and the Israelites) to the Romans. I have learned in all these areas from my peers and fellow scholars. I have taught students ranging from beginners to graduates and often to interested lay people. I have two doctorates in my subject area; I have taught at the university level for forty years, both in the USA and the UK.

Interest in the questions of biblical creation stories is not only the result of my career. Even though I have spent most of my professional years grappling with ancient texts, I almost became a scientist. In my youth I was fascinated by science of all kinds, especially paleontology. In high school, my math and science grades could easily have secured for me a place at one of the major universities. In 1963, I attended the Summer Science Program in Biology for High Ability Secondary School Students at the University of Texas in Austin. Such programs were very competitive. If I remember correctly, the thirty-five participants chosen for this program were drawn from a pool of ten times as many applicants. It was a wonderful experience and very formative for me. As I was being groomed in science, my scientific interest leaned toward the area of biochemistry. I remember believing that this was the scientific area where I would most likely be able to make my mark. I had the good fortune to work in an organic chemistry lab with doctoral students and professors.

One thing made me stick out among this group of stellar young minds. I was the only Bible-believing fundamentalist among them. As a product of the Bible Belt, I was a staunch antievolutionist. Although I knew that scientists generally accepted evolution, I was saddened to find that not a single one of my fellow students had any doubts about the truth of evolution.

Surprisingly, the director of the program at the time indulged me by granting me an evening session in which I could argue my case against evolution. In an hour and a half, I did my best to combine arguments from science, philosophy, and biblical interpretation. In those days I had accepted as true a form of the gap theory (also called old-earth creationism). The gap theory maintained that a large gap in time lay between the original creation of Genesis 1:1 and the state of chaos described in Genesis 1:2 and following. This opinion allowed for an old earth. I thought — along with others who accept the gap theory — this would resolve the conflict with standard scientific conclusions that date the earth in billions of years. I also believed that this gap theory would explain the existence of certain living things on the earth before the creation of the animals and humanity in the Genesis account. Among gap theorists, there were some differences of opinion. Still, most felt that the dinosaurs had lived long before humans. Many believed that dinosaurs had been destroyed by a Luciferian rebellion against God. With all the learning I could muster, I appealed to Isaiah 14 and other biblical passages to harmonize these with some of the findings of science. As with many fundamentalists of the time, I argued that the flood of Noah also helped explain some of the geological and biological phenomena observed in mainstream science. At the end of my presentation, I rejected evolution while building my argument on "scientific" grounds. For me, then, evolution was ultimately unscientific; scientific observation failed to confirm it.

Of course, I convinced no one. There was a certain grudging admiration for my passion and willingness to stand up in such a learned company to defend my views. The experience made it clear to me how difficult it would be to maintain my beliefs if I became a paleontologist. I came to accept that my love for paleontology could not be pursued professionally. I followed a calling I considered to be above any scientific calling, indeed what I thought to be the highest calling. I set out to become a minister.

My basic dilemma about a career was resolved when I was accepted to a Bible college. I went off to study the Bible and studied all the course offerings as intensely as I could. My love of science led me to a course in geology. At the time, many Bible colleges used the book *The Genesis Flood* by John Whitcombe and Henry Morris.[1] It had achieved a subcultural place of prominence among Bible colleges. Having been interested in and having read about paleontology from a young age, I finally was given a "biblical perspective" with a veneer of science to weigh alongside the views of paleontologists. Whitcombe and Morris seemed to be up to the task of answering scientists who supported evolution.

My youthful solution, the gap theory, could not resolve all the problems of squaring the Bible and paleontology. Because I accepted the biblical story of Adam and Eve as accurate historical events, I accepted the related belief that all human history had to be compressed into 6,000 years of biblical and postbiblical history. For those with a scientific bent, this is no easy proposition. Consider the following reconstruction of human history that would have to follow. The Egyptian history was thought to begin no later than about 3000 BCE (before the Common Era), and the history of Mesopotamia also had its beginnings about the same time. According to the figures of the Hebrew Bible, Noah's flood wiped out the world in year 1656 AM (*anno mundi*, or year of the world; i.e., 1,656 years after creation). According to this biblical calculation, it was about 350 years until we meet Abraham, who is often dated as appearing on the scene at around 2000 BCE. Basic arithmetic would then place the creation narrative in Genesis 1:2 at around 4000 BCE. That would mean that the flood would have come during the history of both Egypt and Mesopotamia, at around 2350 BCE. The simple reality is that there survive enough ancient records to form a reasonable understanding of the history of both Egypt and Mesopotamia around this time. There is no evidence in these sources or the archaeological data for a worldwide flood.[2]

1. John C. Whitcombe and Henry M. Morris, *The Genesis Flood: The Biblical Record and Its Scientific Implications* (Phillipsburg, NJ: Presbyterian and Reformed Publishing, 1961).

2. For a time I was very much taken with the ideas of Immanuel Velikovsky (1895–1979), who, in his book *Ages in Chaos: A Reconstruction of Ancient History from the Exodus to King Akhnaton* (Garden City, NY: Doubleday, 1952), attempted to rewrite ancient history by lowering many events in Egypt by about five hundred years. For example, he argued that Queen Hatshepsut (usually dated about 1500 BCE) was the Queen of Sheba, and the Land of Punt was Palestine. He also argued that the Ipuwer Papyrus — *Admonitions of an Egyptian Sage* (normally dated to the Egyptian Old Kingdom or late in the second millennium BCE) — described the exodus of the Israelites from Egypt. However, I soon found that no Egyptologist, even those who are conservative evangelicals, accepted Velikovsky's ideas. They simply could not be supported from the data.

Although a true believer, I did not become a minister. Instead, I made a shift to academic life as a biblical scholar. After earning a bachelor of arts, I did church work (including answering Bible questions by mail and writing articles for denominational magazines) and also began graduate work. My studies included learning the languages of Scripture: Greek, Hebrew, and Aramaic. Eventually, my first job as a university teacher was to teach Greek and Hebrew. It was during this time that I wrote some articles against evolution and also articles in defense of the Bible against biblical "critics" in denominational magazines. I was fervent in my belief and confident in the truth of the Bible. I came to appreciate the importance of studying the cultural background of the Bible. I sought to understand the context in which the biblical texts had been written and formed, the world of the Bible. What could the history and literature of the ancient Near East – the very cradle of the Hebrew Bible (the Old Testament) – teach me about Scripture? This was the world in which the people of ancient Israel had emerged and lived. Likewise, what could the Greco-Roman world – the very world in which Jesus and the New Testament writers had lived and worked – teach me about understanding the New Testament? The desire to obtain proper knowledge on these subjects drove me to pursue a doctorate. My search eventually led me to the School of Theology at Claremont (California). I enrolled in the PhD program in religion at the Claremont Graduate University.

Even though I specialized in studying the Hebrew Bible (Old Testament), I managed to take some New Testament courses as well. It was important for me to study early Jewish literature and history that related to the Bible. This included the Dead Sea Scrolls. One of my academic mentors, William Brownlee, had been in Jerusalem when the Scrolls were first discovered. He helped to unroll part of the first lot so they could be photographed.[3] At Claremont, I had the opportunity to work with the actual texts of the Bible, accessing ancient copies of biblical manuscripts and by studying how ancient Jewish sources understood the books that had become our Bible. The ultimate question for me was plain: Where did the very letters and words of the Bible come from? I was curious about the study of "textual criticism." This

3. These had been sold to Mar Samuel, who was head of a Syrian Christian monastery in Jerusalem, and he later sold them on to scholars. For a fuller account of the discovery of the Dead Sea Scrolls, see James C. VanderKam, *The Dead Sea Scrolls Today*, 2nd ed. (Grand Rapids: Eerdmans, 2010).

branch of study examines the manuscript remains in Hebrew and Greek to establish the earliest recoverable readings. As a young scholar, this is what I really wanted to know.

I also had the opportunity to study a variety of other languages that pertained to the Bible: Ugaritic, Akkadian,[4] Aramaic (some parts of the Old Testament are written in Aramaic), and other Semitic languages. The context and languages of the Bible fascinated me endlessly. I had wanted to take Egyptian and was able to study some Coptic.[5] I studied Greek translations of the Old Testament, sometimes called the Septuagint because of the legend of seventy Jewish scholars who were once believed to have translated the Hebrew as early as the third century BCE.

It was a chance to read literature that was written contemporary with, or even earlier than, the biblical text. I was reading the writings, in translation or in the original languages, written by people who were part of the world in which the ancient Israelites lived and worshiped. In time, I taught courses in ancient Jewish literature and the New Testament.

I was very skeptical of what my various professors told me. It became my vocation to do my own research and reach my own conclusions. As a conservative Christian early in my studies, I learned very well what was widely believed in scholarship. I often sought for ways to argue against it. However, knowledge of the various civilizations in the ancient Near East and the ability to read the Bible in the original languages changed my perception. It was an eye-opener for me, without a doubt. Though hard work, these studies were also an exhilarating experience, a lifelong intellectual expedition of exploration and discovery that continues to this day. I hope to share with you in the pages that follow how this rich tapestry that undergirds the Bible brings to light exciting possibilities for how to read the creation narratives today.

4. This was a language of Assyria and Babylonia that scholars study by reading the cuneiform clay tablets archaeologists have unearthed for us.

5. Coptic is the latest stage of the Egyptian language, a northern Afro-Asiatic language spoken in Egypt until at least the seventeenth century. For much of its history, the Egyptian language was written in hieroglyphs or a script form based on hieroglyphs (such as the Demotic script). However, Coptic was written in a modified form of the Greek alphabet.

2

CREATION IN THE BIBLE

The Ancient Near East and the Beginnings of the World

When studying the Bible in the original languages, I came across things I had never seen before. Passages that I had read in the past suddenly took on new significance. Often enough, these discoveries did not fit with my beliefs. In studying the literary context of the Bible (in particular, ancient Near Eastern and early Jewish literature), the additional information often contradicted several cherished beliefs. Some of these tensions could be reconciled with the way I had looked at the Bible, but in other cases, it was difficult to integrate the new data. The next few chapters will unpack some of these tensions.

The Creation Account in Genesis 1

The first verse of the Bible is often translated: "In the beginning God created the heaven and the earth." The initial sentence of the Bible describes the universe as geocentric; that is, the earth is the center of all things. The earth is separate from "heaven" (or "the heavens"; the Hebrew word is almost always plural in form); it is not part of heaven, which in turn is quite unlike the cosmos as we know it today.[1] Heaven is seen from the perspective of the earth. It is the earth that is, as the Hebrew tells us, formless and in chaos. The earth is covered in darkness (Gen 1:2). The earth concerns the author

1. It is sometimes stated in nonspecialist literature that it is a dual. This is incorrect. Neither is there a plural meaning but, like the English word "heavens," it refers to a single entity.

of Genesis. The vast universe, with its billions of galaxies and trillions of stars, was far beyond the author's ken.

Genesis 1:3 does not begin to describe the creation of the universe in a logical, scientific sequence. The cosmos in Genesis lacks anything that relates to the big bang, expansion, novas, neutron stars, or even a solar system with planets circling the sun. There is nothing here that demonstrates that the author of Genesis has any knowledge or consciousness of the rest of the universe. Rather, in brilliant and wonderful Hebrew prose, the text plays on the theme of "darkness" in Genesis 1:1: "'Let there be light,' and there was light." The expression in Hebrew is simple, yet graphic.

The language is evocative and meaningful, almost poetic. But from a modern scientific point of view, Genesis 1:3–5 does not make sense. There is "light" but no light source. Light is "separated" from darkness, even though we know that darkness is simply the absence of light. Yet the light becomes "day"; and the darkness "night," followed by something even more bizarre: "there was evening, and there was morning – day one." There is not yet any sun, moon, or stars. Where did "evening" come from? Where did "morning" come from? There is no sunset that marks off the evening; there is no sunrise that ushers in morning. How can it be a "day"? And where is the rest of the universe in all this?

Genesis 1:6–8 deepens the mystery. Now God makes a "firmament." What is a "firmament"? Today's readers come across the word "firmament" and understandably tend to think of the sky as understood by modern scientific method. Even in English, the word "firmament" implies something "firm," a solid and rigid surface. That is also the implication of the Hebrew word here translated "firmament." Indeed, the same Hebrew word can even refer to a bowl or metal basin. A similar word is used of a hammered metal plate in Numbers 16:38 (Num 17:3 in Hebrew). The verb form can refer to hammering metal to a thin plate (Exod 39:3; Num 16:39 [Num 17:4 in Hebrew]; Jer 10:9). In Job 37:18, the author uses the verb form of "firmament" to paint a picture of the clouds or heavens being stretched out like a metal mirror.[2] This is also the way that early Jewish commentators understood the Hebrew text. An ancient commentary on Genesis states: "When the Holy

2. A related Phoenician word refers to a golden bowl. The use of this language was widespread in the biblical world.

One, blessed be He, ordered, 'Let there be a firmament in the midst of the waters,' the middle layer of water solidified, and the nether heavens and the uppermost heavens were formed . . . on the second day it congealed; thus 'Let there be a firmament' means 'Let the firmament be made strong.'"[3] There follows a discussion about how thick the firmament was; some say it was two fingers thick, but others say it was many miles.

Just as the English word "firmament" implies something "firm," so the Hebrew word here implies something of actual substance. The Hebrew author is not simply talking about an atmosphere; he is communicating that something *solid* was created that keeps the waters "underneath" from joining the waters "above." This "firmament" – this inverted bowl covering the earth – is called "heaven." Once again, there is the puzzling "evening and morning, a second day."

In Genesis 1:9–13 God gathers all the waters into one place, to make up the "seas," letting the dry land appear, called "earth." While the word "seas" is in the plural, there is nothing about continents. This is a poetic and stylized description of creation. Then plants are commanded to sprout from the earth. What sort of plants? Ferns and cycads? Horsetails? Ancient vegetation we know from fossils? The author talks only of plants "seeding seeds" and "fruit trees bearing fruit with seeds." The text ignores or is not concerned with other sorts of plants. Also, there is the inexplicable fact that plants are in place, yet there is still no sun to provide the energy needed for plant growth. Finally, there is once again the enigmatic "evening and morning, a third day," even though the sun necessary to mark off these times of day still does not exist. While not a scientific description, it is a beautiful piece of literature with appealing imagery, narrative simplicity, great use of repetition, and dramatic storytelling.

In Genesis 1:14–19 God commands "lights" to appear in the firmament. These lights are "to separate light from darkness." This is astonishing in view of Genesis 1:4–5, where God is already said to "separate light from darkness." The lights are never named as "sun" and "moon" here, although the terms are frequently used elsewhere in the Hebrew Bible. Why? The

3. Gen. Rabba 4:2, translation from Harry Freedman and Maurice Simon, trans. and ed., *The Midrash Rabbah: Genesis*, 3rd ed., 2 vols. (London: Soncino Press, 1977).

answer seems to be that the sun and moon were both gods among the surrounding peoples. So in Genesis, they are merely created lights and not gods in their own right. Finally, on the fourth day, we have the heavenly bodies necessary for a proper evening and morning, night and day. But where are these heavenly bodies? Modern readers assume that Genesis is describing the sun around which the earth circulates and the moon that rotates around the earth. That is not the case in Genesis. The language suggests that sun and moon are set in this firmament and thus move across the solid sky, the inverted basin holding back waters above it. Belatedly, the "stars" are also mentioned as being created alongside the "greater light" and the "lesser light" (1:16). The language is wonderfully poetic while being scientifically vacuous. Although the light has been around since day one, only on the fourth day are the sources of light created. The universe described is not even Ptolemaic with earth as its center,[4] much less Copernican with the sun at the center of the solar system.[5] What is expressed in stunning Hebrew is a description of the universe that is similar to other cosmological descriptions in the literature of the ancient Near East.

On the fifth day (Gen 1:20–23) God creates the creatures of the waters and the creatures of the air. Among the sea creatures created are the "great creatures of the sea" (NIV). Readers might infer that the text refers to large sea creatures such as whales. However, given the context of ancient Near Eastern imagery, the mention of these "great creatures" is probably not merely an evocative phrase for large sea life. In other texts (both biblical and in the broader literature of the time) the "great creatures of the sea" are some monsters of chaos that fight against the deity. We will look at this more closely below. Finally, on the sixth day, all the land creatures are created, including humans.

A new twist comes in Genesis 1:29–30. The text follows that moving passage (vv. 26–28) where it is said that men and women are to be created in the image of God. Now the author describes a peaceable food chain: from the lofty humans in the image of God to the lowliest animal, humans,

4. "Ptolemaic" refers to the second-century geographer and astronomer Ptolemy of Alexandria and his belief that the earth is at the center of the universe with the sun, moon, and planets revolving around it.

5. "Copernican" refers to the mathematician Copernicus (fifteenth century) and his understanding that the planets — including the earth — revolve in orbits around the sun.

beasts, and fowl; all are to feed on plants. Genesis is even more precise and indicates that they are to feed specifically on seed-bearing plants (Gen 1:29). Whatever theological meaning may be deduced, it is patently clear that this is neither science nor history. The digestive systems of lions, leopards, and even domesticated cats are not designed for green plants. The balance of nature, in fact, is such that rabbits would overrun the countryside without meat-eating foxes and weasels. Without wolves and lynxes, deer would eat up their food supply until they starved to death. Today, we are very conscious of the importance of the "balance of nature." However, if one reads Genesis 1:29–30 as literally true, there would have been no balance of nature even in the Edenic state. The implication of such a reading is that – in spite of the carnivorous digestive system that God had created them with – carnivores were meant to eat plants, which they are not equipped to process.

It will not do to suggest some method to reconcile modern biological science with a literalistic reading of Genesis. Why do lions have large bone-crunching teeth and jaws and sharp claws? To catch turnips? To subdue and kill rutabagas? Lions were never meant to compete with wildebeests for the grass of the African savannah. Rather, they are majestically built to keep down the numbers of wildebeests and other herbivores. This helps prevent much larger numbers of these animals from starving to death. Genesis does not describe reality; it describes an Edenic utopia. The language of Genesis 1 is compatible with the language of Isaiah 2:2–4 and Micah 4:1–3, which also paint a utopian picture but this time of a future peaceable world that is a complete break from the present all-too-violent world. The future world is actually contrary to nature in the same way that the world of Genesis 1 is contrary to the physical world – to nature – as the Hebrews would otherwise have observed it, let alone the world we understand in the light of modern science. To further evidence the vegetarianism of Genesis 1 as being a theological trope, consider this: according to Genesis 4:2–4, Abel brought offerings from his flock to God. Offerings involved killing the animal and burning portions of its carcass on the altar, including the "fat portions" (4:4). Normally, the one who made the sacrifice ate the rest of the animal that remained after the altar portions were burned. In other words, killing animals for food and other purposes was part of the author's world. However, in Genesis 1 the author described the world in a way that was far removed from the actual world around him. The reason has to do with the purpose of the text within the narrative, its literary and theological intent.

The story in Genesis 1 is brilliant. Perhaps the language is best described as "heightened prose," since it does not take the form found in passages more clearly poetic by the conventions of Hebrew poetry, though admittedly, there is still a lot not fully understood about ancient Hebrew poetry. The language, the expression, and the overall message are the work of a Hebrew literary genius. The text is not the writing of a scientist in any modern sense. The sooner we jettison our categories when we approach the text, the better we can appreciate it on its own terms. There is nothing in the chapter that suggests any advanced knowledge of cosmology, astronomy, geology, biology, or physical reality. As we have seen, the author's understanding of astronomy has the earth with a basin covering it in which are embedded the sun, the moon, and stars that somehow travel or move with the sphere of the disk of the earth. The writer knows nothing of solar systems, galaxies, red giants, white dwarfs, interstellar clouds, or supernovas.

The author is describing creation from her or his perspective. Out of historical necessity, this is a perspective shaped by the understanding of the universe and the world of the people of the ancient Near East in that time. The writer had no special modern scientific knowledge of the universe, or the big bang, or the age of the earth, or the development of life over billions of years. The author is very much a part of the culture in the ancient Near East. What makes the text of Genesis unique for us is the Hebrew encounter with God that occurs within that thought world. The takeaway is the theological insight and belief that God somehow and sometime created the world. The text's exalted and formalistic prose suggests a primordial past beyond the memory of living cultures. However, it is described in the literary context of the ancient Near East, the contemporary literary context of the author of Genesis. The physical details of the mechanics of how God created were unknown to him. Given the highly evocative language of Genesis 1, they seem not even to have mattered to the author. It was enough that God said, and it was! The faith of the author is expressed in the highly graphic, colorful, and metaphorical account that opens the Bible. The value of Genesis 1 is theological, and for nurturing the life of faith. It did not ever express itself as empirical science. To read it that way is to do violence to this passage of Scripture. The text of Genesis 1 and the world of its author belong to the ancient Near East.

The Significance of the Babylonian Creation Epic

In the nineteenth century, a rich array of ancient Babylonian texts was discovered. One of the most interesting among these is the Enuma Elish, sometimes referred to as the Babylonian creation epic. The Babylonian creation epic is one among those found at the Library of Ashurbanipal at Nineveh.[6] The hero in this telling is Marduk, the city god of Babylon. This is the version that scholars have access to today, but there is evidence that there were other versions. For example, we know the Assyrians had a related creation story in which the leading figure is the god Ashur, the chief god of the Assyrians. The original story likely goes back at least to the late second millennium BCE, if not earlier.

The Babylonian epic tells of how the world of the gods was threatened by a chaos monster named Tiamat. The gods look for a champion. Marduk volunteers, on the condition that he would be made king of the gods. The gods agree and Marduk and his allies fight and defeat Tiamat. He then uses her body, which was mostly water, to create the universe. Tiamat is split in two. Marduk uses one-half to form the sky and the other half to form the earth. From the Enuma Elish:

> Valiant Marduk . . . made firm his hold over the captured gods,
> Then turned back to Tiamat whom he had captured.
> The Lord tramples upon the frame of Tiamat,
> With his merciless mace he crushed her skull. . . .
> He calmed down. Then the Lord was inspecting her carcass,
> That he might divide(?) the monstrous lump and fashion artful
> things.
> He split her in two, like a fish for drying,

6. The find is named after Ashurbanipal, the last great king of the Neo-Assyrian Empire. It is a collection of thousands of clay tablets; all manner of texts from Mesopotamia from the seventh century BCE.

Half of her he set up and made as a cover, (like) heaven.
He stretched out the hide and assigned watchmen,
And ordered them not to let her waters escape.[7]

Even in this fantastic creation narrative, we find that there was a mass of waters above the sky and another mass below the earth, from the two halves of Tiamat's watery body.

When the Enuma Elish was first discovered, scholars recognized an astonishing parallel to the account in Genesis 1. This may not seem so at first because the biblical account and the Enuma Elish are also different. Babylonian mythology undergirds the Enuma Elish. In Genesis, creation unfolds not through a cosmic battle with the forces of chaos. Rather, creation is simply composed of inert elements in the hands of God who creates. God does not do battle with the forces of chaos.

However, there are some deep connections between the two creation stories. In Genesis we find a highly skilled and knowledgeable writer – indeed, a capable polemicist – who has written an account of creation in which God is completely sovereign. This is an obvious contrast to the Babylonian accounts. A careful reading will reveal that the Babylonian epic influenced the text of Genesis. Many scholars have argued that Genesis 1 came together during the Neo-Babylonian period, a period of Mesopotamian history that began about 626 BCE and ended in 539 BCE. This is a time when Babylonian traditions and literature would have been especially available to Jewish writers. The Babylonian Enuma Elish and Genesis 1 share common language. A mere coincidence is not likely. The Hebrew word for "deep" is from the same root as the name of the chaos monster Tiamat in the Babylonian epic.

Thus the very language of Genesis 1 suggests that the active demonic opponents of the Babylonian story have become inert elements that are mere tools in God's hands. Rather than facing in combat a powerful monstrous foe in Tiamat, the God of Genesis 1 acts on the "deep," a place of lifeless stuff from which creation is begun. In the Babylonian story, Tiamat's watery body is separated, while in Genesis there is a separation between the waters and the waters (Gen 1:6). In Genesis 1:21 we meet great sea animals that we will

7. Translation of Benjamin R. Foster, *Before the Muses: An Anthology of Akkadian Literature*, 2 vols. (Bethesda, MD: CDL Press, 1993), 376–77.

discuss briefly below. These are not monsters of chaos like Tiamat or as they appear in other stories from this period in the ancient Near East.

The writer of Genesis 1 has turned the Babylonian version on its head. God does not fight active enemies as Marduk does but shapes and molds natural elements to produce an ordered cosmos. The Hebrew writer has deliberately told a creation story that asserts that the Hebrew God, YHWH, is sovereign over the gods of Babylonian mythology, which here become no gods at all. The forces of chaos have become lifeless elements that God shapes as he will. As one Old Testament scholar put it, creation is no longer a battle but a "job of work."[8]

This is only part of the story. There are other biblical passages that give a different perspective on creation.

Other Creation Passages

It is important to recognize that different creation imagery is used elsewhere in the Bible. Genesis 1 describes creation in a particular way, using particular language to tell a particular story. The passage makes a number of religious and moral points. But in other passages creation is described differently, and the story of creation takes a different form. None of them can be said to be scientific, but they all express something about God and his relation to mankind, the world, and perhaps even in some cases the universe.

For example, while Genesis 1 has little reference to "heaven" or "the heavens" beyond references to the firmament, in a couple of places we find the rather poetic expression, "He stretches out the heavens like a tent curtain" (Ps 104:2; cf. Isa 40:22). This is poetic language and will not be bent to fit our modern scientific notions of sky or space. In Psalm 104:5, a few verses later, we read the earth is set on "foundations" that make it firm. This again reflects ancient Near Eastern cosmology (which will be described in more detail below). Here the earth is made firm upon the waters by means of "foundations."

8. See John Day, *God's Conflict with the Dragon and the Sea* (Washington, DC: Catholic University of America Press, 1985).

The problem is that some of the connotations in these passages might be lost on someone reading the Bible in English (or other modern translations). Familiarity with the Hebrew language and the ancient Near Eastern background illuminates the text. With the grammatical and literary context of the biblical texts more clearly in view, a remarkable scenario unfolds. Archaeological discoveries made in the 1920s and 1930s offer new light on familiar biblical texts. One particular site provides a rich yield, a site in northern Syria called Ras Shamra: the ancient name of the city proved to be Ugarit. In 1928, texts were discovered there. They were written in an unknown script and language. The language was quickly deciphered; it was a Northwest Semitic language closely related to Hebrew. Many words are cognate or linguistically related. They seem to have the same origin in earlier Semitic as the Hebrew words found in the Bible and often look like an archaic form of Hebrew. The two languages have many words in common. Some of the most common words are the same. Other words are the same, though perhaps used more or less frequently. Sometimes a similar or related word will have a different range of meaning in the two languages.

The Ugaritic mythic texts relate in poetic form the doings of various gods. Among the more important of this divine cast of characters is the head of the pantheon, El (which is the same name used, at times, for the God of Israel), and the dynamic storm god Baal, well known from some biblical texts. There is also El's consort, Asherah, and Anat, who is sister and consort of Baal. Anat is found only a couple of times in biblical passages (e.g., Judg 3:31; 5:6), but Asherah often appears (as a plural word).

The Ugaritic texts drew attention to other creation stories as well as conflict narratives between the deity and forces of chaos. Although the texts from Ugarit had some similarities, they were quite different from the Mesopotamian creation stories. The Ugaritic stories found parallels in various biblical creation passages but not primarily in Genesis 1. Although these biblical texts had been available to Bible readers, scholars had not realized their full significance until they had other ancient Near Eastern texts to compare with them. With these different texts in clear view, it became obvious that the ancient Israelites had stories about God that shared deep similarities with those that circulated among their Near Eastern neighbors in Syria and Mesopotamia. For example, Psalm 74:13–14:

You [God] drove back "Sea" in your strength,
You broke the heads of the large sea creatures upon the waters,
You shattered the heads of Leviathan,
You gave him as food for the inhabitants of the wilderness.

The Hebrew word translated "large sea creatures," or "whales" in some contexts, was the same word that was originally used for monsters of chaos that threatened to overwhelm the world, the monsters that had to be defeated by God. Leviathan is one of those chaos monsters. In fact, "Sea" – although the word in Hebrew can refer to the ocean – in this context is also threatening God and his creation, just as it does in some of the ancient Near Eastern accounts. Read this way, God is not attacking his creation – the ocean or poor whales or other large animals. This would make no sense. Rather he is doing here what Marduk or Baal or other gods do in other ancient Near Eastern stories of creation or primordial stories. They battle to defeat the forces of chaos, giant monstrous creatures and the ocean personified as an opponent to the order of the gods, and God does the same.

The next passage is similar, Psalm 89:6–11 (89:7–11 in Hebrew):

The heavens praise your wonder, O YHWH,[9]
Your faithfulness in the assembly of the Holy Ones.
For who in the heavens is like YHWH,
Who can compare with YHWH among the Sons of the Gods?
El is greatly feared in the Council of the Holy Ones
Feared by all those around him.
YHWH, God of Hosts, who is mighty like you, Yah?[10]
You rule the swelling of the Sea,
You still the surgings of its waves,
You crushed Rahab like a corpse,
With your strong arm you scattered your enemies.

9. In the traditional Hebrew text, the divine name YHWH was written without vowel points. Although Hebrew scholars suggest the name was pronounced something like Yahweh, this is not absolutely certain. Thus, I have written just the consonants of the word.

10. "Yah" is a shortened form of YHWH.

The book of Job also has some interesting passages. The first is Job 26:12–13:

> By his power he stills the Sea,
> By his skill he struck down Rahab,
> By his wind the heavens were calmed,
> His hand pierced the fleeing serpent.

Once again, the God of Israel fights against monsters of chaos that threaten the divinely created order. "Sea" again here is not the watery ocean but rather an active divine opponent. The same can be said of Rahab. This is a name and creature unique to Israel since the name is not attested in any other ancient Near Eastern literature. As such, it is hard to say too much about Rahab except that the word is used in the company of the other chaos monsters in such passages. The "fleeing serpent" is an expression that has parallels in Ugaritic texts in which the storm deity defeats Leviathan, who is also called the "fleeing serpent."

Similarly, Job 40–41 describes two creatures, known as Behemoth and Leviathan. Some commentaries identify these with living natural creatures, such as the elephant and the crocodile. Some less wisely have even suggested dinosaurs! But the description in Job does not fit any natural animals, alive or extinct. Simply put, no living animal breathes fire (41:11–13)! While even in biblical times all animals have been prey to humans, these two are beyond human capture, subduing, or killing (40:24–41:1). In the text of Job, they are understood to be supernatural creatures that exist in a world between the world of nature and the divine world. According to later Jewish tradition, these two creatures would become the centerpiece of the messianic banquet. One of the most graphic passages is from Isaiah 27:1:

> In that day YHWH will punish with his sword
> — great, cruel, and mighty —
> Leviathan, the fleeing serpent,
> Leviathan, the twisting serpent.
> He will slay the *large sea creature* which is in the sea.

Once more, it is obvious that YHWH is not making war on innocent animals. Like a warrior with a sword, he is fighting the monsters of disorder that threaten to overwhelm the ordered creation. Here Leviathan is pictured as a "serpent." Interestingly, some of the same words are used here that occur in the Ugaritic texts. The following Ugaritic passages shed some light on the language from Isaiah. The god Baal describes past exploits:

> Did I not destroy (Sea) the darling of El,
> did I not make an end of (River) the great god?
> Was not the dragon captured and vanquished?
> I did destroy the wriggling serpent,
> the tyrant with seven heads.[11]

A similar set of events is ascribed to another deity a bit later on:

> . . . for all that you smote Leviathan the slippery serpent
> and made an end of the wriggling serpent . . . ,
> the tyrant with seven heads.[12]

The language used in the Hebrew Bible and in the Ugaritic sources is strikingly similar. "Leviathan the slippery [or "fleeing"] serpent" and "Leviathan the twisting serpent." Scholars acquainted with both Hebrew and the Ugaritic language and literature can easily see that the usage is too close to be accidental. The "large sea creatures" are also found in the Ugaritic text, as is Leviathan, along with the "slippery/fleeing serpent" and "twisting serpent."

When we look at the Ugaritic texts from northern Syria, we discover texts that have many contact points with the Hebrew Bible. The religion expressed in these ancient Ugaritic texts inhabits a conceptual world similar to the worship of YHWH in ancient Israel. A passage from Isaiah will make this even clearer.

11. KTU 1.33.37–39; Manfried Dietrich, Oswald Loretz, and Joaquín Sanmartín, eds., *Die keilalphabetischen Texte aus Ugarit, Ras Ibn Hani und anderen Orten/The Cuneiform Alphabetic Texts from Ugarit, Ras Ibn Hani and Other Places*, 3rd ed. (Münster: Ugarit-Verlag, 2013). This translation is from John C. L. Gibson and Godfrey R. Driver, *Canaanite Myths and Legends*, 2nd ed. (Edinburgh: T&T Clark, 1978), 50. See also 2 Bar 29:4, which reads: "And they will be nourishment for all who are left."

12. KTU 1.5.1.1–3; translation from Gibson and Driver, *Canaanite Myths and Legends*, 68. See also KTU 1.5.1.27–30; translation in Gibson and Driver, *Canaanite Myths and Legends*, 69.

Isaiah 51:9–11a combines imagery of the age-old battle between YHWH and the chaos monsters with the exodus and other imagery:

> Awake, awake, clothe yourself with strength,
> O arm of YHWH,
> Awake as in days of old,
> As in generations of long ago.
> Did it not hack Rahab,
> Piercing the sea creature?
> Did it not dry up Sea, the waters of the Great Deep?
> He established the abysses of Sea as a way for the redeemed to go over.
> And the redeemed of YHWH will return and come to Zion with rejoicing.

In a passage composed with great skill, the poet has combined, within the space of a few lines, references to YHWH's defeat of the chaos monsters (Rahab, the sea creature, the Sea itself, and the deep). He weaves the references to these ancient foes into the story of the crossing of the sea at the exodus from Egypt and ties the whole into a hope for the future exodus in which God's people will return from their captivity in foreign lands to Jerusalem. For all this creativity, the writer was working with traditional material. The defeat of the forces of disorder and chaos by God was a part of this tradition.

As we have seen, a number of passages from the Hebrew Bible offer a rather different creation story. In Genesis 1, the story is a brilliant description of the "work" of God in forming the earth. In some of the psalms, in Job, and in Isaiah, we found something much less polished and rather closer in tone and substance to some of the ancient Near Eastern myths. We read of the God of Israel doing battle with living creatures that represent the forces of chaos. In these passages the imagery is quite different from Genesis 1. God is a warrior who battles monsters such as Sea, Rahab, the great sea creature, and Leviathan the many-headed serpent who sometimes twists and sometimes flees. Some of these biblical passages are very similar to the story of the battle between Marduk and Tiamat in the Enuma Elish or between Baal and various monsters in the Ugaritic myths. In one particular Ugaritic tale, the storm god fights and defeats "Sea." Unfortunately, the tablets with the ending of the story are partially destroyed so we cannot be sure exactly how this battle ends in the Ugaritic sources.

Similar to Genesis 1, these additional biblical narratives describe God's activity in creation but offer no scientific insight. Indeed, as with all biblical creation accounts, they are a poetic or metaphorical way of expressing a theological truth. They are not descriptions of geology, biology, or astronomy. They reveal the nature of divinity.

Conclusions

Divine creation is a major theme in the Bible. Many biblical passages are about God establishing his rule over the world and over nature. We have examined but a few of these passages. Even this brief sampling has demonstrated the rich variety of concepts and language in these narratives as they describe creation. The God of Israel as revealed in these passages is bringing his order and design to the world. This idea dominates the biblical text beyond the key passages in the early chapters of Genesis. It is the specific subject of many biblical passages.

Genesis 1 is an anomaly. The act of creation is a "job of work." God examines the chaos before him and begins to put it right, like a farmer creating a garden out of a waste patch of land. However, other biblical passages draw on a way of describing God and creation that is much older: one that has parallels in the literature of neighboring ancient Near Eastern peoples. Just as the Babylonian god Marduk defeats Tiamat and creates the universe from her body, and the Ugaritic god Baal defeats "Sea," so YHWH defeats the monsters of chaos, Leviathan, Rahab, and "Sea." These ways of talking about God's work in creating the world were part and parcel of the thought world of the biblical writers. However, there were other approaches.

The writer of Genesis 1 makes a sophisticated theological move, now describing creation as God doing a "piece of work." The primordial forces of the deep, "Sea," and the great sea creature – usually depicted as battling YHWH – are now simply inert elements that YHWH shapes into the cosmos. God starts with the scene of chaos ("formless" and "empty" as the Hebrew is rendered from Gen 1:2). However, instead of battling these realities, he shapes them into an ordered universe, in six days. As the Hebrew Bible is organized into its current form, we have an exciting and moving version of the creation story at the beginning. However, this was not the original version of the creation telling in ancient Israel. The older story is the one that resembles those of the other nations in the ancient Near East, where YHWH battles and defeats various monsters of chaos and gives birth to creation. This more ancient story is alluded to in some texts of the Bible, as we have seen, but the full mythology is not unpacked. We get an inkling of the complete story from the texts of neighboring peoples such as those at Ugarit. We may never know the full story behind the creation narratives in the Bible.

Genesis 1 is not a sketch unveiling a view later intuited by modern science. The story is told from a geocentric point of view; the earth is at the center of the cosmos. As noted above, there is no big bang, no vast expanse of galaxies, and no heliocentric solar system in Genesis. The heavenly bodies — the sun, moon, and stars — are described as lights in the inverted dome of the "firmament." Light exists days before there are any heavenly bodies. It is a beautiful piece of heightened prose, a magnificent piece of literature that uses structure and symbol to undergird the importance of divine creation. The astonishing greatness of the cosmological bodies and the forces that operate in and through them have no bearing in this passage. In fact, nothing in modern scientific understanding at all has any bearing on the text. Rather, we have an extended metaphor of God as a divine builder, sorting out the disorder of a building site as told in the unfolding days of creation.

The author of Genesis 1 seems especially concerned to counter the creation narrative based on the divine battle found elsewhere in the Hebrew Bible and especially elsewhere in the ancient Near East. Whereas God is often pictured as fighting supernatural foes that want to bring disorder into the cosmos, the writer of Genesis 1 turns these chaos monsters into inert elements that are like clay in the hands of a divine potter. Sea and light and firmament and beasts do not fight or resist their creator. God simply inserts them into the world he is constructing and arranges them according to his will. This way, the other forces are marginalized; the writer is asserting monotheism and the uniqueness of the Israelite God.

We must, as careful biblical readers, abandon the notion that the writer of Genesis 1 is painting a vision of the universe that is commensurate with a scientific model of creation. The knowledge of the cosmos that would enable such parallels was not in existence at the time. The knowledge and understanding of people in the ancient Near East are the conceptual bricks and mortar for the descriptions found in the Bible. The cosmology, astronomy, geology, and biology of Genesis 1 – insofar as these terms even make sense in this context – did not extend beyond what was known by other learned people of the time (see the next chapter). The writer was writing theology; he was giving us a religious and spiritual account that answered the question, "Who are we?" The scientific picture is available to believers today, but not in the pages of the Bible.

3

THE FLOOD STORY

Mesopotamian Parallels

One of the enlightening discoveries for me during my doctoral work was the great repertoire of texts from the nations surrounding ancient Israel, especially the cuneiform texts from Mesopotamia to the east and Ugarit to the north.[1] I was surprised and intrigued by their many parallels to familiar biblical texts. We discussed at length the parallels to the creation narratives in the prior chapter. The present chapter will look at another series of texts that parallel material in the book of Genesis: the story of Noah's flood.

Mesopotamian Flood Accounts

When I began the study of the Akkadian language and cuneiform literature, I had long been aware that there were other flood stories in the literature from the ancient Near East. A close look at these stories is quite eye-opening. A recent examination of many of these accounts is given by a scholar at the British Museum, Irving Finkel.[2] We can begin with one of the more famous of these other flood stories, a story in the Epic of Gilgamesh.

1. Cuneiform script is one of the most ancient known systems of writing. It is characterized by unique wedge-shaped marks on clay tablets. Scribes used a blunt reed for a stylus. The word "cuneiform" simply means "wedge shaped" and is derived from the Latin *cuneus* "wedge" and *forma* "shape."

2. Irving Finkel, *The Ark before Noah: Decoding the Story of the Flood* (London: Hodder & Stoughton, 2014).

The Epic of Gilgamesh

Gilgamesh was the hero of a long epic poem that is considered the Mesopotamian equivalent of a Homeric epic, although the Epic of Gilgamesh is older. The best-known version comes from the first millennium BCE, but its roots are much older, very likely originating in the third millennium BCE. The hero of this epic poem was on a quest for immortality. In this story, the survivor of the flood (called here Utnapishtim) had been made immortal by the gods. Therefore, Gilgamesh sought him out because he might have the secret to eternal life. The journey was an arduous and dangerous one, but Gilgamesh eventually reached the abode of Utnapishtim and his wife. Utnapishtim told him the story of the flood, which makes up a good piece of the epic poem.

In Utnapishtim's tale, humans had been created to help the gods and to do the hard labor that the gods did not want to do – building dikes as well as digging and maintaining canals, essential for producing food in Mesopotamia. But as humans multiplied and filled the land, they were "too noisy" and disturbed the sleep of the chief god, Enlil, who decided that he would destroy the humans created by the gods. This sounds a rather trite reason to kill all the people: because they were "noisy." However, some scholars have interpreted the Assyrian text to imply something more serious. Some scholars see a close parallel to the corruption and violence of Genesis 6:11. The gods were forced to swear an oath that they would not warn the humans of the disaster approaching. But Enki, the god of wisdom who had helped create humankind, found a loophole. He did not tell Utnapishtim what was to happen, but instead he spoke to the wall beside Utnapishtim. Enki told the wall about the plans to send a flood and gave instructions on how to escape the flood. The text of Tablet XI reads as follows:[3]

[XI 23] "O man of Shuruppak, son of Ubar-Tutu,
demolish the house, and build a boat!
Abandon wealth, and seek survival!
Spurn property, save life!
Take on board the boat all living things' seed!

3. The Arabic numerals are line numbers.

[XI 28] The boat you will build,
her dimensions all shall be equal:
her length and breadth shall be the same,
cover her with a roof, like the Ocean Below."

. . . .

[XI 81] [Everything I owned] I loaded aboard:
all the silver I owned I loaded aboard,
all the gold I owned I loaded aboard,
all the living creatures I had I loaded aboard,
I set on board all my kith and kin,
the beasts of the field, the creatures of the wild,
and members of every skill and craft.

[XI 87] The time which the Sun God appointed –
"In the morning he will send you a shower of bread-cakes,
and in the evening a torrent of wheat.
Go into the boat, seal your hatch!" –

. . . .

[XI 128] For six days and [seven] nights,
there blew the wind, the downpour,
the gale, the Deluge, it flattened the land.

[XI 130] But the seventh day when it came,
the gale relented, the Deluge ended.
The ocean grew calm, that had thrashed like a woman in labor,
the tempest grew still, the Deluge ended.

. . . .

[XI 148] I brought out a dove, I let it loose:
off went the dove but then it returned,
there was no place to land, so back it came to me.
[XI 151] I brought out a swallow, I let it loose:
off went the swallow but then it returned,
there was no place to land, so back it came to me.
[XI 154] I brought out a raven, I let it loose:
off went the raven, it saw the waters receding,
finding food, bowing and bobbing, it did not return to me.

> [XI 157] I brought out an offering, to the four winds made sacrifice,
> incense I placed on the peak of the mountain.[4]

The Babylonian Priest Berossus

When the flood story in the Epic of Gilgamesh was first discovered in 1853, its remarkable resemblance to the flood story of Genesis was commented upon. The obvious question posed was: Which came first? Scholars agreed that the story in the ancient Epic of Gilgamesh was older than the one in Genesis, by perhaps half a millennium. Another version of the flood story was already known to exist, the Berossus version of the flood story. This telling also has some remarkable similarities to the story of Noah in Genesis. Berossus was a Babylonian priest who wrote an account of Babylonian myth and history in Greek about 300 BCE, shortly after the Greek conquest of Mesopotamia. The version of Berossus had been known for many centuries because the historian Eusebius quoted it in the late third century CE. The natural assumption was that the Berossus text had been influenced by the Genesis account. However, the Mesopotamian text and subsequent study have convinced most scholars that Eusebius quoted faithfully the text that had come to him, and his version probably represents closely the text that Berossus himself wrote — though we may not have Berossus's full account preserved.[5] Most scholars have concluded that the Berossus text is later than Genesis. Careful study, however, has shown that his later text drew on early cuneiform Mesopotamian sources. That is, although the translation is late, the sources were probably much earlier. The problem is that we do not know precisely which specific ancient Mesopotamian texts were used by the priest Berossus. Here is the main account of Berossus:

> 1. [Cronus] appeared to Xisouthros[6] in a dream and revealed that on the fifteenth day of the month Daisios [May] mankind would be destroyed

4. Translation from Andrew George, trans., *The Epic of Gilgamesh: The Babylonian Epic Poem and Other Texts in Akkadian and Sumerian* (New York: Barnes & Noble, 1999), 89–94.

5. See especially the study of Stanley Mayer Burstein, *The* Babyloniaca *of Berossus* (Malibu: Undena Publications, 1978).

6. This is apparently the Greek form of Ziusudra, which is the Sumerian equivalent of "Noah" mentioned in the next text.

by a flood. . . . Then, he should build a boat and embark on it with his kin and his closest friends. Food and drink should be placed in it. He was to load into it also the winged and four-footed creatures and to make everything ready to sail. . . . he built a boat five stades [c. 3,000 feet] in length and two stades [c. 1,200 feet] in breadth. . . .

2. [On the third day] after the flood had come and swiftly receded, Xisouthros released some of the birds [to determine if they might see somewhere land which had arisen from the waters]. But finding neither food nor a place on which to alight, the birds returned to the ship. After a few days Xisouthros again released the birds and these again returned to the ship but with their feet covered with mud. On being released a third time, they did not again return to the ship. Xisouthros understood that land had reappeared. Tearing apart a portion of the seams and seeing that the boat had landed on a mountain, he disembarked. . . . It is also said that the land in which they found themselves was Armenia.[7]

Sumerian Flood Story

Scholars of Mesopotamia (called Assyriologists) found other earlier versions of the flood. Some of these stories are not only much older but also are evidently the ancestors of the Epic of Gilgamesh and Genesis. This literary trail is documentary evidence for the development of the flood story in Mesopotamia, going back to the early Sumerian tradition that is at least as old as the third millennium BCE.

One of the first known references to the flood is found in the Sumerian King List, about 2400 BCE. The text states: "The Flood swept thereover. After the Flood had swept thereover, when the kingship was lowered from heaven the kingship was in Kish" (lines 39–42).[8] There is also a Sumerian version of the flood story in what is often referred to as the Eridu Genesis.[9]

7. Burstein, *The* Babyloniaca *of Berossus*, 20–21.

8. Thorkild Jacobsen, *The Sumerian King List*, AS 11 (Chicago: University of Chicago Press, 1939), 140–41.

9. The translation on the following page follows Wilfred G. Lambert and Alan R. Millard, *Atra-ḫasīs: The Babylonian Story of the Flood* (Oxford: Clarendon, 1989), 143–44. Cf. William Wolfgang Hallo, ed., *The Context of Scripture*, 3 vols. (Leiden: Brill, 1997–2002), 1:513–15.

The clay tablets that preserve this story were copied about 1600 BCE, but the story itself probably originated earlier, certainly much earlier than the Epic of Gilgamesh. The Sumerian flood hero's name is Ziusudra (the Greek name Xisouthros is a form of this):

151 . . . a wall
152 Ziusudra hea[rd], standing by its side,
153 He stood at the left of the side-wall [. . .]
154 "Side-wall, I want to talk to you [hold on] to my word,
155 [Pay atten]tion to my instructions:
156 On all dwellings (?), over the capitals the storm will [sweep].
157 The destruction of the descent of mankind [. . .]"

. . . .

201 All the destructive winds (and) gales were present,
202 The storm swept over the capitals.
203 After the storm had swept the country for seven days and seven nights
204–5 And the destructive wind had rocked the huge boat in the high water,
206 The Sun came out, illuminating the earth and the sky.
207 Ziusudra made an opening in the huge boat.
208 And the Sun with its rays entered the huge boat.
209 The king Ziusudra
210 Prostrated himself before the Sun God,
211 The king slaughtered a large number of bulls and sheep.
(gap in the text)

. . . .

256 (Who) gave him life, like a god,
257 Elevated him to eternal life, like a god.

Here part of the story is missing because we have only some damaged tablets of it. Much of the story of preparing the ark and of the flood itself is lost. Yet, remains of the tale are remarkably similar both to the Epic of Gilgamesh and to the biblical story.

Atrahasis

Another flood story can be found in our sources, somewhere between the Sumerian version above and the Gilgamesh versions. This narrative is known by its Akkadian title, Atrahasis; it is named after its hero. This version is very important for understanding of the development of this story because it is preserved in a near-complete form and is also older than any part of the biblical text. The story is similar to the Sumerian flood story but is much longer and we find elements that are much closer to the form of the story we find in Gilgamesh.[10]

(Enki speaks to Atra-hasis by addressing the wall of the house:)
20 "Wall, listen to me!
21 Reed wall, observe all my words!
22 Destroy your house, build a boat,
23 Spurn property and save life.
. . . .
25 The boat which you build
[gap]
29 Roof it over like the deep.
30 So that the sun shall not see inside it
31 Let it be roofed over above and below.
32 The tackle should be very strong,
33 Let the pitch be tough, and so give (the boat) strength.
34 I will rain down upon you here
35 An abundance of birds, a profusion of fishes."
. . . .
[Tablet III, ii]
32 Clean (animals) [. . . . .]
33 Fat (animals) [. . . . .]
34 He caught [and put on board]
35 The winged [birds of] the heavens.
36 The cattle (?) [. . . . .]

10. The translation given here is from Miguel Civil in Wilfred G. Lambert and Alan R. Millard, *Atra-Ḫasīs: The Babylonian Story of the Flood* (Oxford: Clarendon Press, 1969), 138–72.

37 The wild [creatures (?) [. . . . .]
38 [. . . . .] he put on board
. . .
42 . . .] he sent his family on board,
(The coming of the flood is described)
. . .
[Tablet III, iv]
24 For seven days and seven nights
25 Came the deluge, the storm, [the flood]
. . .
[Tablet III, v]
34 [The gods sniffed] the smell,
35 They gathered [like flies] over the offering.

Comparing all the versions of the flood story, we can see that, despite some differences in details, the structural similarities are striking. The version closest in time to the story in Genesis, the Epic of Gilgamesh, is the one that seems to be most like the story of Noah, especially in regard to the details about the birds being sent out. The majority of biblical scholars believe that the flood story in Genesis has its literary origins in Mesopotamia. The Mesopotamian flood story was taken up by Hebrew scribes and adapted to Hebrew conventions and theology.

Table 1: Resemblances between Genesis and Other Ancient Flood Stories

Genesis 6–9	Gilgamesh XI	Berossus	Atrahasis	Sumerian Account
Humans are wicked	Humans rebellious			[Humans rebellious]
God warns Noah	Ea warns Utnapishtim	Cronus warns Xisouthros	Enki warns Atrahasis	Enki warns Ziusudra
Noah builds ark	Utnapishtim builds boat		Atrahasis builds boat	[Ziusudra builds boat]

Genesis 6–9	Gilgamesh XI	Berossus	Atrahasis	Sumerian Account
Ark 300 × 50 × 30 cubits	Dimensions equal (a cube?) 120 × 120 cubits	Boat: 5 × 2 stades		
Takes two of every kind plus family	Takes all living creatures, kin, and craftspersons	Takes winged and four-footed creatures, kin, closest friends	Takes various creatures and family	
Provisions ark	Bread and wheat	Food and drink loaded	Provisions provided	
Forty days/nights of rain	Seven days of rain		Seven days and nights of rain	Seven days and nights of rain
Fountains of deep and windows of heaven				
Raven sent out and does not return; dove sent and returns; dove sent and returns with olive leaf; dove sent and does not return	Dove sent and returns; swallow sent and returns; raven sent and does not return	Birds sent and return; birds sent and return with mud on feet; birds sent and do not return		
Ark lands in mountains of Ararat		Ark lands in Armenia		Ark lands
Offers sacrifices	Offers sacrifices		Offers sacrifices	Offers sacrifices

It may come as a surprise to some that biblical authors borrowed material from Mesopotamia. However, this is not unusual. Ancient Near Eastern and Greco-Roman cultures profoundly influenced the various authors of the Bible. These authors lived and breathed within particular cultural contexts. They used language and ideas in their literary and religious environments to tell the story of YHWH and Israel.

Hebrew Cosmology

A basic understanding of Hebrew cosmology illuminates the account of the flood in Genesis. How did the ancient Hebrews picture the universe? How was it constructed? What were its parts? One example will illustrate this point: Some commentators debate the amount of water needed for a worldwide flood as described in Genesis 6–9. Some have even attempted to explain the amount of water as natural phenomena. This approach encounters numerous problems. With a literal reading of the flood, we are left wondering where the water needed to cover Mount Everest has gone. Such a feat would require far more water than what is found in all the oceans today, about three times more.

Ancient cosmology helps to explain the story of the flood. Genesis 7:11 tells us (see similarly 8:2):

> All the springs of the great deep burst open,
> And the hatches of the heavens were opened.

Let us go back to the creation story in Genesis 1 for a moment where similar language is used. In Genesis 1:6–8 we read:

> And God said, "Let there be a firmament in the midst of the waters, and let it divide between the waters and the waters." And God made the firmament, and it divided between the waters which are under the firmament and the waters which are above the firmament. And it was so. And God called the firmament Heavens, and there was evening and there was morning, the second day.

Genesis 7:11 now becomes clear. The Hebrew word translated "springs" does indeed refer to "springs" or "water sources" (even "fountains"). However, the reference to the "great deep" uses a Hebrew word that may refer to the ocean. The ocean is sometimes alluded to as the "deep" in Hebrew, but as we saw in chapter 2 the phrase the "great deep" points to a mass of water on which the earth floats. It refers to the "waters below" the earth (Deut 4:18). In the flood these seem to break past their normal boundaries and wash over the earth like a tidal wave. "Firmament" in early Hebrew referred to a solid object. It was not a metaphorical word that paralleled a modern understanding of the sky or atmosphere. Rather, the ancient Hebrews conceived of it as a solid covering, a hemisphere or bowl, over the world that had a mass of waters above it. As for the "hatches" of the heavens or sky, in Hebrew they can refer to a hole in a surface, like a "smoke hole" or a "chimney" or even a "window" or some other opening. The picture given to us is of the "waters above the firmament" pouring out of these holes or openings like a flood from above, to add to the torrent of water welling up from below the earth. This was no ordinary rainstorm; it was a return to the watery chaos out of which the earth had been created in Genesis 1:6–8.

The story in Genesis does not refer to endless days of conventional rainfall, but an opening in the firmament pouring out the water from above even as water from the great deep burst upward. Now, let us take a larger view at the universe as these ancients picture it, as one artist has conceived it. (See The Ancient Hebrew Conception of the Universe on Plate 1.)

The universe that the writer of Genesis understands is a three- or four-tiered cosmos. This picture is commonly found in other ancient Near Eastern sources. The earth is overarched with a bowl-like sky. Above the sky is a mass of waters, held back by the solid barrier of the sky or firmament. There are windows or hatches through which precipitation comes. Normally, this happens in a controlled fashion. The earth also is afloat on a mass of waters. These waters are held in check. This was a widespread view of the world in the ancient Near East and eastern Mediterranean at the time.

In chapter 2 we already took a closer look at the Babylonian creation epic, which pictures the creation of the cosmos as coming about by splitting the watery corpse of the goddess Tiamat. Half of her remains were used to form the sky and the other half to form the earth.[11] The image, like those in Genesis, is one where above the firmament is a huge quantity of water, and the same is true for below the earth. These waters are kept in check until the time of the flood, during which water from above the sky and below the earth is released to inundate the inhabited world.

According to Egyptian cosmology, the world originated in water. In Egyptian thought specifically, the earth was pictured "as a flat, rimmed dish resting upon water."[12] As with all ancient sources, there is some variation in the Mesopotamian concept of cosmology in the texts that survive from the period. However, the general contours of this ancient cosmology seem to remain fairly stable, resembling those we saw from the Enuma Elish. One text, The Bilingual Creation of the World by Marduk, offers a condensed version of the creation of the universe.[13] The world, which begins in water, is created as a raft floating on the waters:

> Marduk wove a raft on the face of the waters.
> He created dirt and poured it on the raft.

This is consistent with the Egyptian texts mentioned above. The cosmology of the ancient Hebrews, reflected in the pages of the Bible, originated among neighbors who articulated this view.

While this image of the cosmos may sound primitive and naive to us in the twenty-first century, the understanding of the cosmos in these texts makes sense of what the ancients observed around them. The Greek philosopher Thales, sometimes regarded as the world's first scientist, embraced this view of the universe. Living in the sixth century BCE, Thales was among a group of early scientists who lived in Asia Minor. In the sixth and fifth centuries BCE a sort of intellectual renaissance took place. We find evidence of a larger number of thinkers and writers in this period, such as Herodotus

11. Enuma Elish IV 135–40.

12. So state Geoffrey S. Kirk, John E. Raven, and Malcolm Schofield, eds., *The Presocratic Philosophers: A Critical History with a Selection of Texts*, 2nd ed. (Cambridge: Cambridge University Press, 1984), 92, though the exact Egyptian text being referred to is unclear, and I have been unable to locate it.

13. The text is partially published in Wayne Horowitz, *Mesopotamian Cosmic Geography* (Winona Lake, IN: Eisenbrauns, 1998), 130–31.

and the pre-Socratic philosophers. Some have used the expression "Ionian enlightenment" because of the flourishing of original thought in a variety of areas of knowledge at this time.

Although Thales's writings are preserved only in part, he was one of the earliest genuine scientists. He is alleged to have predicted an eclipse of the sun in 585 BCE. He argued that everything originated in water; exactly what he meant is debated by historians of Greek philosophy. The Greeks had various philosophical debates about the essential building blocks that made up the endless variety of the external world. However, his writings do seem to be an attempt to describe the world scientifically. According to Aristotle, who knew Thales's work, "Others say that the earth rests on water. For this is the most ancient account we have received, which they say was given by Thales the Milesian, that it stays in place through floating like a log or some other such thing."[14]

If the cosmos or world was afloat on water and under a dome of water as envisaged by Thales and the Mesopotamians (perhaps also the Egyptians), might it not also be the model the writer of Genesis had in mind? This makes good sense of the language in Genesis and explains well how this ancient biblical author imagined and described how the flood took place. It was not merely a heavy rain, whether of seven days or forty days, which would hardly flood the region, much less the world. Rather, the primordial waters that surrounded the world – the waters above the firmament and the waters below the earth – burst through the barriers that normally held them back and engulfed the land. This was not a "normal" flood caused by too much rain; it was a return to the primeval chaos from which the earth originally arose.

Later Jewish authors explain the cosmos in ways that are equally peculiar. For example, the Astronomical Book, a part of 1 Enoch (c. 300 BCE), describes the movements of the heavenly bodies, including the sun and moon.[15] The text makes an effort to approximate the rise and movements of the sun and moon over the course of a year in numerical terms. Despite advances in scientific knowledge, this Jewish text does not demonstrate an awareness of the sophisticated Babylonian astronomy of the Persian or

14. This is from Aristotle, *De caelo* B3, 294a28, translation quoted from Kirk, Raven, and Schofield, *The Presocratic Philosophers*, 89.

15. See 1 Enoch 72–82. For background on the book, see George W. E. Nickelsburg, *Jewish Literature between the Bible and the Mishnah*, 2nd ed. (Minneapolis: Fortress, 2005), 43–53.

Seleucid-Parthian period.[16] One Jewish text explains the movements of the sun through the seasons this way:

> This is the first law of the luminaries: the luminary (called) the sun has its emergence through the heavenly gates in the east and its setting through the western gates of the sky. I saw six gates through which the sun emerges and six gates through which the sun sets. The moon rises and sets in those gates and the leaders of the stars with the ones they lead, six in the east and six in the west, all of them – one directly after the other. There were many windows on the right and left of those gates.[17]

The full passage attempts to describe the movement of the sun (and also the moon) from north to south and back again during the course of the year, which it describes graphically moving through a series of six gates. This description may be an honest attempt to reckon what earthly observers can discern; it is also stylized according to a predetermined scheme and only loosely corresponds to the real movements of heavenly bodies. Other ancient Near Eastern cultures, such as in Mesopotamia, demonstrated far more advanced astronomical knowledge, and the movement of the heavenly bodies was far better understood at this time.[18]

Job 26:7: An Outlier of Ancient Hebrew Cosmology?

The writer of Job has a remarkable knowledge of nature, but that knowledge is rooted in *his own time*. It does not square with modern scientific knowledge. The list of the wonders of nature paraded in Job 38–40 reflects ancient wisdom and knowledge. A number of key passages illustrate the author's understanding of the cosmos. Note first of all the statements about the cosmos found in Job 26:

16. Cf. Matthew Black, *The Book of Enoch or I Enoch: A New English Edition*, Studia in Veteris Testamenti Pseudepigrapha 7 (Leiden: Brill, 1985), 387.

17. 1 Enoch 72:2–3, quoted from the translation of James C. VanderKam in George W. E. Nickelsburg and James C. VanderKam, *1 Enoch 2: A Commentary on the Book of 1 Enoch, Chapters 37–82*, Hermeneia (Minneapolis: Fortress, 2012), 416.

18. See further Nickelsburg and VanderKam, *1 Enoch* 2, 335–407.

[26:5] The shades tremble
Beneath the waters and their inhabitants.
[26:6] Sheol is naked before him;
And there is no covering for Abaddon.

Here the dead are pictured as trembling in fear before God beneath the waters, presumably the cosmic waters on which the earth floats. Underneath the earth is Sheol/Abaddon. In the Old Testament, this was a sort of underworld or the place where the dead normally reside. In some traditions Sheol is associated with mud, mire, and damp (cf. Pss 40:3; 69:3).[19]

[26:10] He drew a circle on the surface of the waters,
At the extreme where light and darkness meet.

There are some variations in the manuscript copies in the Hebrew of this verse. Many scholars think it refers to the earth as a circle or disk on the waters, with the borders of the disk as the boundaries between light and darkness. The language is similar to that of Isaiah 40:22, which states that God sits on the circle or disk of the earth. Proverbs 8:27 speaks of God's cutting a circle on the face of the primal deep. However, according to Isaiah 40:22 God is seated above the circle of the earth. Psalm 24:2 tells us that the earth was founded on waters. All these passages depict the earth as a disk established and floating on the primal waters. What is more, the earth seems to rest on pillars or foundations that rest on these waters, with the heavens stretched over it like a tent, also held up by pillars:

[26:11] The pillars of heaven tremble,
Astounded at his blast.

Other passages in Job, such as 9:6, also talk about the "pillars" on which the earth is founded: the verse tells us the earth has foundations. The Hebrew text here refers to the foundation walls or bases on which the rest of the structure is built. These are laid by God according to Job 38:4. The earth has pedestals, the bases on which pillars rest. Then chapter 26 describes God

19. For example, the Ugartic texts KTU 1.4.8.11–12; 1.5.1.6–8..

fighting the monsters of chaos (26:12–13) – Sea, Rahab, and the Twisting Serpent.

Other passages in Job (along with parallel passages in other parts of the Bible) speak of similar architectural supports of the earth, with similar words such as a cornerstone (38:6) and corners (38:13).[20]

Turning back to Job 26, let us take a look at verse 7. The following is a very literal translation of the Hebrew:

> [26:7] The one who stretches out Zaphon over chaos,
> Who suspends the earth upon/over formless matter.

The first line of the verse is rather enigmatic. *Zaphon* is the name of Mount Cassius in Syria. This was the sacred mountain for the people of Ugarit, who believed the gods dwelt there. For the Hebrews, Zaphon might simply be a shorthand for "north" in general. However, even in the Hebrew Bible it could refer to the dwelling place of God (as in Isa 14:13). Sometimes Mount Zaphon stood in for Mount Zion (e.g., Ps 48:2). When it shows up, it often refers to a heavenly dwelling. The Hebrew word here for "chaos" is the same one used in Genesis 1:1 and it is often translated "unformed," "chaotic," or "waste." It refers to the formless matter, the raw stuff with which God began creation. The words that I translate "formless matter" in the second half of Job 26:7 are a rare Hebrew construction, found only here in the Bible. It clearly parallels the meaning of the word translated "chaos." The meaning should start to become clear now. The language in Job 26:7 seems to suggest that the earth is suspended over or, more likely, *upon* the primordial waters of chaos. This is a very close parallel to what we saw above for Genesis 1:9–10 and other related passages. There the dry land lay on the face of the deep.[21]

Job 26:7 is a reference to the time of primordial creation. God is stretching out the heavens over the primordial chaos. This description tracks well

20. Cf. also Job 37:3; 38:13; Isaiah 11:12; Ezekiel 7:2. Note also the "ends of the earth" in Deuteronomy 28:49; Isaiah 5:26; 26:15; 40:28; 41:5, 9; 42:10; 43:6; 48:20; 49:6; Job 28:24; Psalms 46:10(9); 48:11(10); 61:3(2); 65:6(5); Proverbs 17:24.

21. This meaning of the Hebrew word we are discussing is confirmed by the translation of the Job targum, a Jewish translation of the Hebrew text into Aramaic: "He erects the earth over/upon the waters without anything to support it." For the text of the targum, see David M. Stec, *The Text of the Targum of Job: An Introduction and Critical Edition* (Leiden: Brill, 1994), 173.

with God's activity in Genesis 1:6–8 and then resting the earth/land on the primordial waste. The understanding of the cosmos presupposed by the book of Job seems to be in agreement with other biblical passages. This translation of Job 26:7 fits well the cosmological understanding of the rest of Job. However, some translations of Job 26:7 – for example, the King James Version or the New International Version – render the language a bit differently:

> He spreads out the northern skies over empty space;
> He suspends the earth over nothing. (NIV)

Some interpreters read this and leap to the conclusion that it squares with a modern understanding of the cosmos. However, even in the NIV, the language of "chaos" or "formless matter" does not track with modern astronomy. What are these "northern skies"? Sky is only a perception of humans on the earth. In true scientific parlance, there are no "skies," only space and various bodies (stars, planets, comets, asteroids, etc.) within them. There is no "north" in space, and the earth is not suspended at all – it is a body in orbit, going around the sun at a tremendous speed. Those who wish to see a modern understanding of astrophysics in the verse are simply reading into its poetry their own modern notion.

Ibn Ezra was a medieval Jewish commentator known for his pioneering work on Hebrew grammar. In Ibn Ezra's writing on this passage, he describes this part of Job in a way that seems foreign to both the context of Job and our own view of the universe. He interprets the last part of Job 26:7 this way: "for the earth is suspended in the middle of the spheres."[22] What does he mean by "the spheres"? As a medieval scholar would, he understands the earth as being surrounded by, and at the center of, several heavenly spheres. Medieval scholars accepted the Ptolemaic cosmology, according to which the earth is the center of the universe.[23] Surrounding it are a number

22. For a critical text and also a Spanish translation, see Mariano Gómez Aranda, *El Commentario de Abraham ibn Ezra al Libro de Job: Edición crítica, traducción y estudio introductorio*, Consejo superior de investigaciones científicas instituto de filología, Serie A: Literatura Hispano-Hebrea 6 (Madrid: Consejo superior de investigaciones científicas instituto de filología, 2004).

23. The reference to Ptolemaic cosmology is confirmed by the study of Mariano Gómez Aranda, "Aspectos científicos en el comentario de Abraham ibn Ezra al Libro de Job," *Henoch* 23 (2001): 81–96 (esp. 88–89).

of translucent spheres in which are embedded the sun, the moon, and stars. These spheres rotate around the earth. Ibn Ezra interpreted Job according to *his* understanding of the universe, just as some moderns interpret it according to *their* own modern scientific knowledge of the universe. The point is that the temptation to read Job, or any ancient description of the universe, in harmony with our own understanding is a well-worn but hazardous practice. The book of Job itself provides clues that its author saw the earth and the universe in ways that fit the ancient Near Eastern culture of the time.

It should surprise no one that the writer of the flood story in Genesis drew on a picture of the cosmos that does not match our modern scientific understanding. Knowledge of a scientific sort developed by observation, even among the ancient Jews. However, they were not at the forefront of an observational or scientific approach. At the time some among the Greeks recognized that the earth was round and suspended in space. A few Greeks (such as Aristarchus) even argued that the earth went around the sun.[24] In comparison, the Jewish understanding lagged behind. The ancient Jews simply did not have the resources devoted to science. The structure of learned society in ancient Israel was not set up to further the accumulation and dissemination of scientific knowledge.

The biblical accounts of the universe found in Genesis and Job simply reflect the best of Jewish knowledge at the time. A worldwide flood was possible to imagine as it fits into a cosmology where the disk of the earth was afloat and also covered by a universe of waters. Today, we have a better scientific understanding of geology and the physical earth.

Conclusion

What strikes an informed reader about the flood stories of the ancient Near East is how readily the story in Genesis 6–9 seems to fit into the conceptual world of other ancient flood stories, especially those from Mesopotamia. Similarly to the beliefs of the Ionic philosopher Thales, the Mesopotamian

24. Aristarchus of Samos (c. 310–c. 230 BCE) was an ancient Greek astronomer and mathematician who is the first known person to have modeled the universe with the sun at the center and the earth in orbit around it.

creation stories (as well as certain Egyptian accounts) and some of the passages in the Bible (including Noah's flood) picture the earth as a boat or raft or disk floating on the cosmic ocean. Genesis 6–9 has most in common with the much earlier flood stories from Mesopotamia. These resemblances are striking, right down to the sending out of birds to see whether there is dry land. There are, of course, as we have seen, many differences in detail. An especially important point is that the Genesis story is from a monotheistic point of view: there are no competing gods, no swearing of oaths not to tell humans, and no deception of one deity by another. It all takes place under the eye of the one God who decides to spare Noah and his family, along with pairs of living creatures.

What seems clear, as most biblical scholars accept, is that the Hebrew writer has taken over a Mesopotamian story but has rewritten and adapted it to fit a Jewish theological framework. Many of the details from these other flood stories were not essential to the story as told in Genesis. While some details remained with little change (such as the birds), other details were modified or dropped. Much was added. The writer in Genesis found the calendar with a particular chronology of events important to preserve, though the forty days and forty nights of rain were imposed on this for literary effect. Since the specifications for building the ark vary among the different stories in Mesopotamia, it is hardly surprising that they differ in the biblical account as well. The story still retains its patent Mesopotamian form but also fits well within these first chapters of Genesis.

When we look at the story closely, it is obvious that it presents an ancient Hebrew picture of the universe, a particular cosmology. It is not a cosmology that can be verified by modern geology. The earth has waters beneath it, left over from the original creation when cosmic waters were subdued. It has a "firmament," a bowl-like dome across which the heavenly bodies move. Doors in this dome allow precipitation, but the dome as a whole holds back further cosmic waters, the "waters above the firmament." But when the flood comes, the cosmic waters beneath the earth and above the sky burst forth and flood the world. This is why the whole world is submerged after only forty days of rain. The cosmological picture given to us is of a return to the primitive watery chaos that prevailed before creation.

4

"AFTER ITS KIND"

Genetics and Evolution

The phrase "after its/their kind" is found at Genesis 1:11–12, 21, 24–25 in the King James Version. The Hebrew word translated "kind" can mean different things: kind, category, or sort. Some creationists misunderstand and distort the meaning of "kind" to support particular ideas about the Bible and science.

The Meaning of "Kind"

What is the meaning or connotation of the Hebrew word translated "kind" in Hebrew? Some creation scientists assume that "kind" represents a closed category that cannot be transcended. They understand variation and even *evolution* to occur within the kind. However, change cannot go beyond the boundaries of the kind. Microevolution within the kind is accepted, but macroevolution that allows one kind to change over time into another kind is not. With this explanation, all the kinds were individually created by God. Some creation scientists coined the term *baramin*, a word that comes from the Hebrew words for "create" and "kind."[1] Publications on baramin often focus on scientific research and neglect careful treatment of the Bible. As a result, they posit this idea of kind as a biblical

1. See especially Todd Charles Wood et al., "A Refined Baramin Concept," *Occasional Papers of the Baraminology Study Group* 3 (2003): 1–14. A number of creationists are unhappy with the term. See the comments in Todd Charles Wood, "The Current Status of Baraminology," *Creation Research Society Quarterly* 43 (2006): 149–58 (esp. 155–56).

teaching and then proceed to explain its implications without reference to what the Bible actually says.

Several creationist scientists feel that their scientific quest for further refinement of the idea of creation according to kinds is justified.[2] It seems fair to dispute these assertions based on the science that is supposed to uphold them.

Baramin – creation according to kinds – is argued and defined in scientific terms. Yet the idea itself is based on a particular reading of Genesis. If the fundamental starting point of a theory is in error, the thesis itself is likely to be in error. In this case most biblical scholars would say that this interpretation of kind or baramin has no exegetical support in the Bible. Let us look at a few places where this same Hebrew word for "kind" is used in the Bible.

Creation scientists rarely connect kind with species in Linnaean taxonomy.[3] Kind is more often related to genus or even family.

In Genesis the various "grasses," "fruit trees," "birds," "beasts," and "swarming things" reproduce after their kind (1:11–12, 21, 24, 25; 6:20; 7:14). Here we might think that "kind" refers to larger categories, such as a dog kind (*Canis*) and a cat kind (*Felis*). If this were so, it would support those who see kind as a higher classification, an inviolable form that would not evolve. It has long been argued by some creationists that "kind" would generally correspond to a group higher than species, perhaps even family, in scientific classification.[4]

However, Leviticus 11 and Deuteronomy 14, texts that describe the animals that can and cannot be eaten, speak of specific animals with the same Hebrew word for "kind" that we find in the Genesis creation stories.

> These from the birds you shall consider abominable; they shall not be eaten; they are an abomination: the eagle and the vulture and the black vulture and the kite and the falcon after its kind; every raven after its

2. See Wood et al., "A Refined Baramin Concept," 10.

3. The particular form of biological classification (taxonomy) set up by Carl Linnaeus in 1735. In this taxonomy there are three kingdoms, divided into classes, and they, in turn, into orders, families, genera (singular: genus), and species (singular: species), with an additional rank lower than species.

4. Wood, "The Current Status of Baraminology," 149–58. According to Wood, George McCready Price had already equated kinds with family as early as 1924. Wood himself is more circumspect.

> kind, and the ostrich, and the nighthawk and the sea gull; and the hawk after its kind; the little owl, the cormorant, and the great owl, the white owl, the pelican, and the bustard, the stork; the heron after its kind, the hoopoe, and the bat. (Lev 11:13–19 // Deut 14:12–18)

> Also, these you may eat from all the winged swarming creatures which walk on (all) fours, which have jointed legs above their feet to hop on the ground. From these you may eat: the locust after its kind, and the bald locust after its kind, and the cricket after its kind, and the grasshopper after its kind. (Lev 11:21–22)

While many of us would think of a falcon kind as one that encompassed both eagles and hawks, Leviticus 11:21 seems to make them separate. We would also tend to lump crickets, grasshoppers, and locusts together. Yet Leviticus 11:21–22 seems to make them different kinds. Similarly, Genesis 8:6–12 indicates that a raven kind and a dove kind were in the ark. In these cases, "kind" is hardly a scientific category that groups related animals at the level of family. Rather, it seems to function as only a vague and indeterminate expression for sort or variety. These texts do not offer a clear scientific classification of larger animal groupings.

This is a problem for creationists. None of the uses of "kind" in these passages correspond to modern scientific taxonomy. Sometimes animals are named "after their kind," but then other animals have no such designation. There is no indication that one is different from the other. Neither is there any indication of a nested hierarchy of classification as if some named animals are a species, others a genus, others families, others orders, and so on. In other words, there is nothing here that would allow one to classify animals from a scientific point of view.

Someone might ask, "If the interpretation of these biblical texts is not absolutely certain, how can they be used to oppose the interpretation of those who argue for the fixity of living kinds?" But this is the point! There is no certainty of animal types in the text. There simply is not that level of linguistic specificity. Then how much less can the text be argued to mandate that divine creation of individual kinds implies the impossibility of macroevolution? Those who claim the biblical text teaches the fixity of kinds have no linguistic evidence whatsoever to support their view.

The earliest translation of the Hebrew Bible was the early translation in Greek, commonly referred as the Septuagint. The Greek translation of Genesis is too literal to be helpful, simply repeating "after its kind." However, in Leviticus 11 where "after its kind" is used a number of times, with certain birds declared to be unclean for human consumption, the Greek for Leviticus 11:15 reads, "the raven and those like it."[5] It is also worth noting that a number of recent translations do not give the English rendering "after its kind" but translate the Hebrew in a way that gives a rather different sense. Notice the following examples:

New Jewish Publication Society:
Genesis 1:11: "fruit trees of every kind"
Genesis 1:12: "plants of every kind, and trees of every kind"
Genesis 1:25: "God made wild beasts of every kind and cattle of every kind, and all kinds of creeping things of the earth"
Leviticus 11:14–19: "falcons of every variety; all varieties of raven . . . hawks of every variety . . . herons of every variety"

Revised English Bible:
Leviticus 11:14–19: "every kind of falcon; every kind of crow . . . every kind of hawk . . . the various kinds of cormorant"
Leviticus 11:22: "every kind of great locust, every kind of long-headed locust, every kind of green locust, and every kind of desert locust"

New International Version:
Leviticus 11:14–19: "the red kite, any kind of black kite, any kind of raven, the horned owl, the screech owl, the gull, any kind of hawk, the little owl, the cormorant, the great owl, the white owl, the desert owl, the osprey, the stork, any kind of heron, the hoopoe and the bat"

What about the argument: "The Bible says these animals are to reproduce after their kind. How can that allow for evolution?" The fact is that

5. Similar translations are found in verses 14–16, 19, and 22; see also the additional phrase in Genesis 1:11–12.

evolution also assumes that living things reproduce after their kind! Evolution does not hypothesize that a chimp gives birth to a human or that a fish egg hatches out a lizard. On the contrary, each generation normally has offspring that are like their parents, with only minor differences. The fact that the next generation might have mutations when compared to the generation of their parents does not make them different kinds. Only exceptionally are there major differences from one generation to the next. The normal scheme is that changes occur over a long period of time, through countless generations. Only then can one document that the hereditary line has substantially changed. This is true, whether in standard Darwinian gradual change or more recent theories such as punctuated equilibrium. Even in punctuated equilibrium the process of speciation as the result of changes in the environment is only relatively more rapid. The norm is that such change takes many millennia and many thousands of generations to produce creatures that differ significantly from the original generation. Changes that move lines of heredity beyond the species level take even longer. For example, the fascinating diversification we find in the fossil record called the Cambrian explosion took place over millions of years. This relatively rapid appearance of new forms – of most major animal phyla – dated more than 500 million years ago was also accompanied by major diversification of organisms. It happened "overnight" in geologic terms, but this relatively rapid change that led to creatures vastly different from the pre-Cambrian ones still took millions of years to bring about.

Thus, along with the creationists, and even given that the Hebrew cannot bear the weight of the interpretation some foist upon it, science agrees that creatures reproduce after their kind. This is something that we all observe, be it in our own back gardens or at the local farm. As stated in the Bible, this is simply a truism commonly observed, not a statement of genetic determinism. There is nothing in any of the biblical passages referring to kinds that offers a scientific determination that would allow microevolution but deny macroevolution. There is no comment on evolution as such in this scriptural language. Those preserving the Hebrew language and text over the centuries have never given these passages such an interpretation before a handful of modern creationists offered a novel reading of the Bible.

Ancient Jewish Commentaries

Ancient Jewish commentaries on the Bible are particularly helpful in understanding how the Bible was understood in the past. These were written in later stages of the development of the Hebrew language. Some are in Aramaic, a language related to Hebrew, which was also used to write a few sections of the Bible. These commentaries demonstrate how the phrase "after its kind" was understood in the context of other biblical passages, specifically Leviticus 19:19 and Deuteronomy 22:9–11. These latter passages have to do with "mixtures":

> Your domestic animals you shall not mate with diverse kinds; your field you shall not sow with diverse kinds; and cloth of diverse kinds of fabric shall not come upon you. (Lev 19:19)

> You shall not sow your vineyard with diverse kinds, lest you forfeit [to the temple] the whole harvest – the produce of the vineyard whose seed you have sown. You shall not plow with an ox and an ass yoked together. You shall not wear fabric of wool and linen woven together. (Deut 22:9–11)

Not everything in these passages is relevant to our present discussion. The statement that one should not plow with an ox and an ass yoked together could easily be explained as a concern for animal welfare. The prohibition of garments woven from linen and wool may have been a means to separate the priestly class. The garments of priests were of this mixture.

These regulations seem to be aimed at preventing crossbreeding of various animals and plants. One rabbi is quoted in the Talmud as saying that the flood was caused by breeding wild and domestic animals together:

> Said R. Yohanan, "This teaches that [the men of the generation of the flood] made a hybrid match between a domesticated beast and a wild animal, a wild animal and a domesticated beast, and every sort of beast with man and man with every sort of beast."[6]

6. b. Sanh. 108a, translation from Jacob Neusner, ed., *The Babylonian Talmud: Translation and Commentary* (Peabody, MA: Hendrickson, 2006).

A major section of the Mishnah is concerned with such mixtures.[7] The rabbis understood creation to be ordered and hierarchical. Their reading of Genesis understood the types of creation as being brought into existence in increasing importance, from plants, to animals, to human beings. The law in Leviticus according to this section of the Mishnah was to prevent the mixing of the classes established and consecrated at creation.

Thus, early rabbinic texts interpret the regulations of "kinds" in Genesis as relating to the crossing of various fruits and vegetables. They reflect the understanding of the day and do not go beyond the basic farmer's understanding of reproduction. The ancients did not describe things in scientific terms.

Ancient rabbinic Jewish commentators reflect an understanding of Genesis 1 based on their knowledge of the Hebrew language and tradition. Their worldview was much closer to the time and perspective of the biblical writers themselves. The rabbis based their understanding on the scientific knowledge current in their time. They assumed that certain plants would crossbreed if grown near to each other but also that certain ones would not crossbreed. They further assumed that grafting one plant onto another would yield mixed fruits. Modern scientists would come up with quite different rules about what could and could not crossbreed. The rabbis were simply following the biblical writers themselves. Farmers have been crossbreeding plants for centuries.

The point is that none of the early Jewish commentators understood the Hebrew text as limiting the extent to which animals could breed. On the contrary, the whole emphasis was on the prevention of the sort of crossbreeding that was assumed to be *possible*. How does this relate to our previous discussion of kinds in Genesis and the claims that this was a genetically inviolable class?

As noted above, "kind" has no consistent meaning that can be correlated to what is advocated by some creation scientists. "Kind" does not imply a limit on genetic relationships or genetic change. This is hardly surprising; the biblical text is not a scientific textbook. In order to resolve this problem, those advocating the theory of *baraminology* make the point that the idea

7. The Mishnah is a codification of Jewish oral traditions. As such, it is the first major work of rabbinic literature.

of baramin does not correspond to a particular biblical term. Instead, they have now attempted to produce "after its kind" as a scientific concept that can be confirmed or debated in a scientific context.

Baraminology (the theory that animals exist according to kinds that cannot crossbreed or evolve) was invented to support a particular interpretation of the Bible. So this "scientific" theory is based on a construct drawn from the words of the Bible but then has *abandoned* the essential biblical perspective on which it is founded. In other words, the fundamental assumption of baraminology is that the Bible teaches the immutability of the kind – that there is a limit to evolution or change, which is confined to microevolution and does not include macroevolution. But if "kind" has no consistent definition in the biblical data, how can creationists argue that the Bible teaches there can be no macroevolution?

Baraminology combines the gathering of scientific data with ideological presuppositions. Advocates of baraminology have predetermined their scientific investigation by a particular interpretation of ancient texts. Thus, their investigation will always conclude there is no such thing as macroevolution, which begs the question being studied.

Once this theory is assumed to have biblical legitimacy, then errors of a scientific sort are compounded. Several creationists who argue for the theory of baraminology are trying to reconcile evolution insofar as it seems plausible to them with a shortened creationist time span. For example, many creationists now accept that the horse kind evolved in much the way evolutionists have argued for decades. However, given their other assumptions about the Bible and science, these creationists compress the evolution of the horse kind into a few centuries instead of millions of years! There are enormous biological consequences as a result. With such assumptions in play, what does the distinction between microevolution and macroevolution mean? No evolutionary biologist would regard the development of the horse – from *Hyracotherium*[8] to *Equus*[9] – as anything but macroevolution.[10]

8. *Hyracotherium* is an extinct mammal about the size of a fox that had four-toed forelimbs and three-toed hind limbs. Scientists believe they are the earliest ancestors of the modern horse.

9. *Equus* is the name of a genus in the family Equidae, which includes horses, donkeys, and zebras.

10. David P. Cavanaugh, Todd Charles Wood, and Kurt P. Wise, "Fossil Equidae: A Mono-

As a final footnote to the discussion of baraminology, we might consider the work of Phil Senter. He is a biologist at Fayette State University in North Carolina. He grew up accepting the arguments of creationists until studying evolution in high school. This led him recently to test some of the arguments of the baraminologists.[11] The tests themselves are mathematically based statistical calculations used only by some creationists with advanced degrees in biology and are partially based on technical methods developed in conventional biology. Although most biologists do not accept the particular application of this technology in the particular way worked out by baraminologists, Senter decided to apply these techniques just as the authors themselves did. Senter argues that even their own methods affirm the conclusions of paleontology and evolutionary science. Baraminologists have naturally disputed his results, but Senter maintains that their own biological formulae support evolution.

Developments in Genetics

The laws and regulations in the rabbinic Jewish discussions of Genesis 1, Leviticus 19:19, and Deuteronomy 22:9 are far removed from contemporary genetic science. We know today what the ancients did not. DNA is critical to the development of the individual. More fundamentally, DNA is critical to the makeup of all living things that turn out to be genetically related. All living things are influenced by the DNA in their genes — even viruses, which some scientists do not consider fully alive. Not that long ago, living things were classified by biologists by their external characteristics, physical morphology, and the like. Now genetics are taken into account. What is truly remarkable — that is, remarkable only if it were just a coincidence — is that the genetic relationships of most living things closely parallel the evo-

---

baraminic, Stratomorphic Series," in *Proceedings of the Fifth International Conference on Creationism*, ed. Robert L. Ivey (Pittsburgh: Creation Science Fellowship, 2003), 143–53.

11. Phil Senter, "Using Creation Science to Demonstrate Evolution 1: Application of a Creationist Method for Visualizing Gaps in the Fossil Record to a Phylogenetic Study of Coelurosaurian Dinosaurs," *Journal of Evolutionary Biology* 23.8 (2010): 1732–43; Phil Senter, "Using Creation Science to Demonstrate Evolution 2: Morphological Continuity within Dinosauria," *Journal of Evolutionary Biology* 24.10 (2011): 2197–216.

lutionary history as worked out by earlier generations of scientists who did not have the knowledge of genetics to back them up.

This can be illustrated by comparing a human with a fruit fly. The fly is quite different from a human being. The fly is an invertebrate with an exoskeleton, a segmented body, six legs, compound eyes, and a set of wings. Yet humans share genes with fruit flies. Even more remarkable are the discovered Hox and other regulatory genes. Hox genes are important for the development of the various body segments. The fly and the human have the same eight Hox genes.[12] The regulatory genes appear to be controlling genes that activate or keep inactive parts of the genetic material. Most biological "mechanisms" have DNA material that is shared with "earlier" forms of life but remains inactive, perhaps now in the form of "pseudogenes." It will probably remain inactive if it is not a part of the phylogeny (evolutionary history) of the particular plant or animal in question.

Sometimes, however, this DNA can be turned on by a genetic accident or even in the lab, in which case a creature will be born that has characteristics not possessed by others of the species. For example, the humble chicken has much genetic material that evolutionary biologists say it has inherited from its reptilian ancestors. This idea could once be dismissed as evolutionary speculation, but modern genetics now demonstrates that this is indeed the case. When some of this dormant genetic material is turned on, as it has been in genetic experiments, a chicken develops the initial stages of teeth, like those in some reptiles, because the necessary genes have persisted in chicken DNA.[13] In these experiments, certain dormant genes in a chicken embryo were turned on. The chicken started to develop reptile teeth similar to those of a crocodile. No genetic material was added to the chicken's genome. Rather, existing inactive genes were caused to function once again. The genetic material was not able to produce a full set of proper teeth. This

12. A simplified explanation of this can be found in Neil H. Shubin, *Your Inner Fish: The Amazing Discovery of Our 375-Million-Year-Old Ancestor* (New York: Pantheon Books, 2008), 107–10.

13. Experiments done by Professors John Fallon and Mark Ferguson and PhD student Matthew Harris and others at the Universities of Manchester (in England) and Wisconsin have shown this. See Matthew P. Harris et al., "The Development of Archosaurian First-Generation Teeth in a Chicken Mutant," *Current Biology* 16 (2006): 371–77.

is likely because the length of time from the last toothed ancestor to the chicken is considerable. However, the fact that some genes of the chicken's reptilian ancestor remained intact adds considerable corroboration to what the fossils have already told us about these evolutionary lines.[14]

Another recent experiment has carried this demonstration further.[15] In this case, genetic experiments were conducted to see whether bird beaks might revert to dinosaur-like snouts, since the beak is thought to have evolved fairly late in bird evolution from dinosaurs. As the chicken embryo develops, a large patch of cells on the face contains proteins that produce the beak. Scientists suppressed the proteins that produced the bird beak. The result was that instead of a beak, the embryo developed a pair of rounded unfused bones resembling the dinosaur snout. Much experimental work remains to be done, but this suggests that the genes for the earlier dinosaur type of snout remain preserved in the chicken genome.

Genetic manipulation has recently accelerated in various fields of science. The use of genes to develop new crops or crops with particular characteristics is a prominent example. Another example is the development of yeast that will produce vaccines that once required the use of horses or monkeys. The use of genetically modified yeast cells to make vaccines more quickly and less costly – and also without the unfortunate side effects of those produced by inoculating animals – has been a major breakthrough by scientists. The insight that beneficial genetic manipulation can be done in a laboratory setting illustrates the importance of genetics to the development of organisms. Why are so many organisms related in a vast scheme of life at the genetic level such that these scientific gains and experiments are even possible? If each created form was created independently and uniquely by special creation and in a pattern that is inviolable, why do these genetic relationships prove true time and again? One theological possibility worth considering is that God created the laws of genetics and allowed living things to unfold by the inherent potential of the DNA.

14. Jean-Yves Sire, Sidney C. Delgado, and Marc Girondot, "Hen's Teeth with Enamel Cap: From Dream to Impossibility," *BMC Evolutionary Biology* 8.246 (2008).

15. Bhart-Anjan S. Bhullar et al., "A Molecular Mechanism for the Origin of a Key Evolutionary Innovation, the Bird Beak and Palate, Revealed by an Integrative Approach to Major Transitions in Vertebrate History," *Evolution* 69.7 (2015): 1665–77.

The Question of Transitional Forms and Vestigial Organs

If macroevolution took place, we would expect to find transitional fossils, sometimes referred to popularly as "missing links," that connect an earlier form with a more evolved form. Evolutionists usually talk about "transitional features" rather than "transitional forms," because the various features do not evolve at the same rate, but this is a minor point. There should be evidence for a series of fossilized creatures that demonstrate the transition from an earlier form to a later one. It is important to understand what "transitional forms" are. In a fair bit of creationist literature, transitional forms are described as features of creatures that are half-formed or that possess nonfunctioning characteristics. Kurt Wise, a creationist, provides a good example of this misunderstanding and is worth quoting at length:

> None of the stratamorphic intermediates have intermediate structures. Although the entire organism is intermediate in structure, it is the *combination* of structures that is intermediate, not the nature of the structures themselves. Each of these organisms appears to be a fully functional organism full of fully functional structures. *Archaeopteryx*, for example, is thought to be intermediate between reptiles and birds because it has bird structures (e.g., feathers) and reptile structures (e.g., teeth, forelimb claws). Yet the teeth, the claws, the feathers and all other known structures of *Archaeopteryx* appear to be fully functional. The teeth seem fully functional as teeth, the claws as claws, and the feathers as any flight feathers of modern birds. It is merely the *combination* of features that is intermediate, not the structures themselves. Stephen Jay Gould calls the resultant organisms "mosaic forms" or "chimeras." As such they are really no more intermediate than any other member of their group.[16]

The way Wise casts the evidence of the creature known as *Archaeopteryx* is entirely misleading. *Archaeopteryx* is an excellent example for transitional

16. Kurt P. Wise, "The Origin of Life's Major Groups," in *The Creation Hypothesis: Scientific Evidence for an Intelligent Designer*, ed. J. P. Moreland (Downers Grove, IL: InterVarsity Press, 1994), 227. Wise seems to have recognized this and his fundamental error, because an article written a year later, in 1995, omitted such a statement and also accepted that these examples were indeed examples of transitional forms in evolutionary terms.

features. Wise and some other creationists have simply ignored what evolutionary theory actually proposes.

Evolutionary theory does not propose that intermediate features are half-developed, nonworking bodily features that hang off or sit in the body, waiting around for a few more million years until they are able to function. This is a fantasy, not an evolutionary point of view. On the contrary, transitional features are understood to be fully working parts of the viable body of a given life-form. The precise function of a transitional feature might change and develop over time, but the features normally have a place in the life of the creature. The fossil record attests vestigial features that are in the process of disappearing, such as the hind legs on ancient whales.[17] Eventually, such organs cease to have a function, but even as they diminish, they may still have a viable function. For example, the tiny hind limbs of *Basilosaurus,* a genus of early whale that lived forty to thirty-four million years ago, would no longer bear the weight of the creature. However, these limbs might still have played a part in copulation. Still, over time it is seen that these limbs have continued to shrink. In modern whales these limbs are only internal structures where they serve as the anchor for certain muscles (see further on whales in the next section).

These creatures are recognized as transitional forms by two criteria: their form (morphology) and the geological layer in which they were found in the fossil record (stratigraphy). The external features will show minor or major differences and these will occur in earlier or later strata of the geological record. Though a creationist, Wise lists the various types of fossils with transitional forms that might demonstrate macroevolution:[18]

(a) Intermediates within a species, in form and over time
(b) Intermediate species
(c) Intermediates in higher groups (genus, family, and order)
(d) Series of intermediates

17. On vestigial organs and their implication for both evolution and creationism, see Phil Senter, "Vestigial Structures Exist Even within the Creationist Paradigm," *Reports of the National Center for Science Education* 30.4 (2010).

18. Kurt P. Wise, "Towards a Creationist Understanding of 'Transitional Forms,'" *CEN Technical Journal* 9.2 (1995): 218. My list attempts to interpret Wise's list, which is given in more scientific jargon. Wise is trying to find terminology that is more neutral from a creationist perspective, but he has faithfully summarized what evolutionists would expect to find.

Wise points out that the first group (group a) is the most problematic for evolutionary theory. Species seem to have persisted unchanged for long periods of time. That is why, as with all science, evolutionary theory itself has evolved, and evolution is now sometimes described as *punctuated equilibrium* (new species evolve only when there are major changes in the environment). On the other hand, there are many fossil examples of types b through d, which leads a creationist like Wise to admit that these "*intermediates expected by macroevolutionary theory is surely strong evidence for macroevolutionary theory*."[19]

To illustrate transitional forms, we could consider a number of graphic examples. The classic one is the horse.[20] Other examples include fish to tetrapods[21] and early birds from dinosaurs.[22] But for a clear example, we have chosen to look at whales.[23] As this example will demonstrate, paleontologists in recent years have amply demonstrated the transitional features between forms. The remarkable finds in the fossil record continue even as this book goes to press.

19. Wise, "Towards a Creationist Understanding of 'Transitional Forms,'" 219 (italics mine).

20. For example, Cavanaugh et al., "Fossil Equidae," 143–53.

21. A good summary of information is found in the recent book by Jennifer A. Clack, *Gaining Ground: The Origin and Evolution of Tetrapods*, 2nd ed. (Bloomington: Indiana University Press, 2012). I would suggest that anyone who wants to explore the question should begin there. See also Robert Carroll, *The Rise of Amphibians: 365 Million Years of Evolution* (Baltimore: Johns Hopkins University Press, 2009).

22. See especially Philip J. Currie, Eva B. Koppelhus, Martin A. Shugar, and Joanna L. Wright, eds., *Feathered Dragons: Studies on the Transition from Dinosaurs to Birds* (Bloomington: Indiana University Press, 2004); Gareth Dyke and Gary Kaiser, eds., *Living Dinosaurs: The Evolutionary History of Modern Birds* (Chichester: Wiley-Blackwell, 2011); Luis M. Chiappe and Lawrence M. Witmer, eds., *Mesozoic Birds: Above the Heads of Dinosaurs* (Berkeley: University of California Press, 2002).

23. In addition to J. G. M. Thewissen, *The Walking Whales: From Land to Water in Eight Million Years* (Oakland: University of California Press, 2015), the information in this section comes from a variety of sources. See, among others, Carl Zimmer, *At the Water's Edge: Macroevolution and the Transformation of Life* (New York: The Free Press, 1998); Richard Ellis, *Aquagenesis: The Origin and Evolution of Life in the Sea* (New York: Viking, 2001); J. G. M. Thewissen and Sunil Bajpai, "Whale Origins as a Poster Child for Macroevolution," *BioScience* 51.12 (2001): 1037–49; J. G. M. Thewissen and E. M. Williams, "The Early Radiations of Cetacea (Mammalia): Evolutionary Pattern and Developmental Correlations," *Annual Review of Ecology and Systematics* 33 (2002): 73–90; Darren Naish, "Fossils Explained 46: Ancient Toothed Whales," *Geology Today* 20.2 (2004): 72–77; Sunil Bajpai, J. G. M. Thewissen, and Ashok Sahni, "The Origin and Early Evolution of Whales: Macroevolution Documented on the Indian Subcontinent," *Journal of Biosciences* 34.5 (2009): 673–86.

Example of a Transitional Form: Whales

In the original 1859 edition of *The Origin of Species*, Charles Darwin mentioned whales only once, commenting:

> In North America the black bear was seen . . . swimming for hours with widely open mouth, thus catching, like a whale, insects in the water. . . . I can see no difficulty in a race of bears being rendered, by natural selection, more and more aquatic in their structure and habits, with larger and larger mouths, till a creature was produced as monstrous as a whale.[24]

Because this statement was criticized by some reviewers, Darwin dropped it in subsequent editions. In a later edition, he did briefly discuss the possibility that the baleen of baleen whales had evolved from teeth.[25] But as with other lines of creatures, the lack of transitional forms was a major frustration.

Whales, porpoises, and dolphins are "cetaceans" (scientific family Cetacea). Ancient (extinct) whales are referred to as Archaeoceti. Modern whales are separate and divided into two types, the toothed whales (Odontoceti) and baleen whales (Mysticeti). Evolutionary theorists have long argued that whales evolved from a carnivorous, four-legged hoofed land mammal, a view supported by Darwin. William H. Flower, in a speech given in 1883,[26] rejected the popular view at his time that whales had a seal-like ancestor, because whales swim with their tails whereas seals use their hind legs. Flower went on to summarize his own view:

> We may conclude by picturing to ourselves some primitive generalized, marsh-haunting animals with scanty covering of hair like the modern hip-

24. This passage was accessed from the selections of Darwin's writings in James A. Secord, ed., *Charles Darwin, Evolutionary Writings* (Oxford: Oxford University Press, 2008), 181.

25. See Charles Darwin, *The Origin of Species by Means of Natural Selection, or The Preservation of Favoured Races in the Struggle for Life*, 6th ed. (London: John Murray, 1897), 170–73. Fossil finds have produced whales that had teeth along with their baleen; also, in the embryonic development of baleen whales, teeth appear briefly, before being absorbed. Cf. Robert J. Asher, *Evolution and Belief: Confessions of a Religious Paleontologist* (Cambridge: Cambridge University Press, 2012), 130–31.

26. William H. Flower, "On Whales, Past and Present, and Their Probable Origin," in *Essays on Museums and Other Subjects Connected with Natural History* (London: Macmillan, 1898), 209–31.

> popotamus, but with broad, swimming tails and short limbs, omnivorous in their mode of feeding, probably combining water-plants with mussels, worms, and freshwater crustaceans, gradually becoming more and more adapted to fill the void place ready for them on the aquatic side of the borderland on which they dwelt, and so by degrees being modified into dolphin-like creatures inhabiting lakes and rivers, and ultimately finding their way into the ocean.[27]

Flower made his prediction based primarily on a study of living whales. Interestingly, a century before Flower's address, the anatomist John Hunter (1728–1793) noted that "there are numerous points in the visceral organs of the *Cetacea* [whales] which far more resemble those of the *Ungulata* [hoofed mammals] than those of the *Carnivora* [to which seals belong]."[28] He showed that these include the complex stomach, the simple liver, the respiratory organs, and the reproductive organs. Hunter's arguments proved to be astonishingly prophetic, as Flower's speech a century later was to show.

In the late nineteenth century, the fossil whale *Basilosaurus* had been discovered (for a description, see below). By 1970, a century afterward, the only intermediate form that had been found in the evolution of the Cetacea was still *Basilosaurus*, showing how little the fossil evidence for whales had changed. But in more recent years, many more specimens have been discovered and studied. For example, some further *Basilosaurus* skeletons were found with clearly formed legs and feet, which extend about two feet in length, on a sea animal of about sixty feet in length. These feet are clearly too small to support the weight of such creatures on land. Further, they would not have been much of an aid to swimming. It was suggested at the time of discovery that they may have served to facilitate copulation.[29] Beginning in the early 1980s and continuing since, a whole host of intermediate cetacean forms have been unearthed.

While a lot has been made about the lack of hind legs in modern Cetacea, there are actually many additional differences in anatomy between whales and land mammals:

27. Flower, "On Whales, Past and Present, and Their Probable Origin," 231.

28. Flower, "On Whales, Past and Present, and Their Probable Origin," 230.

29. Philip D. Gingerich, B. Holly Smith, and Elwyn L. Simons, "Hind Limbs of Eocene *Basilosaurus*: Evidence of Feet in Whales," *Science* 249 (1990): 154–57.

1. Position of nostrils
2. Reduction of hair
3. Loosening of the spine to allow undulation
4. Shortening of the neck and fusion of the neck bones
5. Loss of hind limbs
6. Forelimbs becoming flippers
7. Teeth morphology
8. Adaptation of hearing
9. Development of echolocation (in toothed whales)

The different anatomies that can be cataloged across these different animals demonstrate their intermediate characteristics. However, this is not a "line of genealogy" – a unilinear progression – as if they descended one from another in a straight line. The precise genealogy between species here (or in any evolutionary pathway) may never be known since the fossil record will never be complete. Still, these examples illustrate the sort of transitional forms and characteristics that existed between land mammals and modern whales for their evolution to be possible. Now we will trace the developments and changes in anatomy from land creature to the aquatic whales as illustrated by the transitional forms currently known from fossil evidence.[30] The most notable feature that changes over time between these various creatures is the hind limb and pelvis. However, this major development is not a feature changing in isolation. Other changes in the body's anatomy from terrestrial to aquatic creature need to be discussed as well.

Pakicetus had four legs, hoofs, and a long tail.[31] Although it apparently spent time in shallow fresh waters, it was very much a terrestrial animal. Why is it classified as a cetacean, then? In spite of its clear terrestrial origins, *Pakicetus* had several anatomical features that allow it to be classified with the whales as a cetacean. Remarkably, the hearing apparatus of *Pakicetus* was anatomically very similar to those in whales – and only whales. Other whale-like

30. Many of these features are summarized in Thewissen and Bajpai, "Whale Origins as a Poster Child for Macroevolution," 1037–49.

31. Sirpa Nummela, S. T. Hussain, and J. G. M. Thewissen, "Cranial Anatomy of Pakicetidae (Cetacea, Mamallia)," *Journal of Vertebrate Paleontology* 26 (2006): 746–59; Sandra I. Madar, "The Postcranial Skeleton of Early Eocene Pakicetid Cetaceans," *Journal of Paleontology* 81.1 (2007): 176–200.

characteristics of *Pakicetus* include the shape of the skull, its thick bones (that helped it to wade or even dive underwater), and its teeth. As we shall see, the hoofs are a persistent feature in later whale forms. (See *Pakicetus* on Plate 2.)

Ambulocetus natans, which appears much more whale-like, shared many features with *Pakicetus*. It was similarly four-legged with a long tail and large back feet that were important in swimming (much like *Pakicetus*). However, the body shape of *Ambulocetus natans* recalls the torpedo shape familiar from more recent cetaceans. The hands and feet still had mobile joints rather than the fixed flipper of modern whales. The digits ended in small hoofs, confirming Flower's argument in the nineteenth century that whales shared kinship with the artiodactyls (even-toed ungulates). The spinal column vertebrae suggest that *Ambulocetus natans* swam using a combination of undulation of the tail and the hind feet, much as river otters do. This mode of locomotion is characteristically intermediate between terrestrial creatures and modern whales (even though shared by otters). They could also move about on land and in shallow water, but their body structure suggests they did not do so rapidly. They had no external ears, but their hearing apparatus suggests they could hear well underwater. (See *Ambulocetus natans* on Plate 3.)

Kutchicetus currently represents the best-preserved specimen of remingtonocetids. This family of cetaceans includes such genera as *Remingtonocetus*, *Dalanistes*,[32] and *Andrewsiphius*.[33] They are characterized by nasal openings near the front of the skull. Remingtonocetids were found near the coastline of modern Pakistan and India during the Eocene era some forty-eight to thirty-eight million years ago. *Kutchicetus* had a narrow, elongated skull that tapered to a long sharp snout with many teeth. Its body shape recalls the otter with a long tail and four relatively short limbs. It was well adapted to swimming, probably with an emphasis on undulating its tail and hind parts like a modern otter. In this it seems to be further developed beyond *Ambulocetus natans* and more closely approximating the locomotion of modern whales. Its short legs would still have allowed it to come out on land. Scientists have used isotope oxygen analysis — a method that examines the distribution of

32. *Dalanistes* were closely related to *Remingtonocetus* with a combination of terrestrial and amphibious adaptations. *Dalanistes* was the larger of the two related species.

33. This early cetacean had an elongated snout higher than it was wide. Small holes on the tip of the snout hint that it may have had whiskers. Its eyes were located midway in the cranium in such a way that it looked a little like a crocodile. It also sported a large crest on the back of its skull.

certain stable isotopes and chemical elements and has widespread applicability in the natural sciences – and found that as the fossil locations suggest, the creature lived in salt water, apparently shallow coastal waters and lagoons. Two further developments seem to push them further and make them more similar to modern whales. *Kutchicetus* had a fat pad underneath the lower jaw that helped it hear when underwater. The balance canals (of the ear) were reduced in size much as in modern cetaceans. (See *Kutchicetus* on Plate 4.)

Rodhocetus is one of the protocetids – another early whale type[34] (along with *Georgiacetus*, *Maiacetus*, and *Artiocetus*).[35] Its body was similar to a whale, with the characteristic streamlined shape. It had small hind limbs, and its front flippers had webbed toes. The question of a tail fluke is difficult to establish because these tissues did not readily fossilize. However, evidence suggests that none of the protocetids had tail flukes, an important characteristic of modern whales. In *Georgiacetus* the size of the neural spines and the fact that the sacral vertebrae of the spine were not fused strongly suggests that the creature swam primarily by means of undulations of its tail and hind limbs.[36] It could have come on land for the purposes of mating and giving birth, as modern seals do. On land, its movement would have been slow, but it hunted in the water. *Georgiacetus* may have been more aquatic than some of the other protocetids. The nostrils of *Georgiacetus* were no longer at the end of the snout but partway back along the head, a transitional feature that resulted in the blowholes of modern whales. In the case of another protocetid, *Maiacetus*, a skeleton was discovered that apparently had a fetus inside.[37] This creature was set to give birth head first, like a land animal (modern whales are born tail first to prevent drowning). The digits of *Maiacetus* bore hoofs like earlier cetaceans, which demonstrate its existence as an amphibian creature. (See *Maiacetus* on Plate 5.)

34. This is a diverse group of cetaceans with specimens discovered in Asia, Europe, Africa, and North America.

35. Philip D. Gingerich et al., "New Whale from the Eocene of Pakistan and the Origin of Cetacean Swimming," *Nature* 368 (1994): 844–47; Philip D. Gingerich et al., "New Protocetid Whale from the Middle Eocene of Pakistan: Birth on Land, Precocial Development, and Sexual Dimorphism," *PloS One* 4.2 (2009): 1–20.

36. *Georgiacetus* is an extinct protocetid from about forty million years ago that ate fish and swam in the Suwannee Current in the coastal sea that once covered the Southeastern United States.

37. This genus of early cetacean was discovered in Pakistan.

Dorudon is another early proto-whale whose shape was more rigid with a torpedo-shaped body. It lived at the same time as *Basilosaurus*, an early whale that lived forty to thirty-four million years ago. It was initially mistaken for a reptile; hence the name, which means "king lizard." *Dorudon* closely resembled *Basilosaurus*. One of the biggest differences between these two ancient whale species is that *Basilosaurus* was a much larger size. *Basilosaurus* was fifty to sixty feet long, compared to twelve to fifteen feet for *Dorudon*. Its body was rather more snake-like. Some paleontologists group *Dorudon* and *Basilosaurus* together, but others are not so sure. Both species demonstrate a further shift of the nostrils toward the back of the head, both sport a tail fluke, and for both the front limbs have become flippers (with immobile wrists). Both still retained tiny hind legs that would not have supported their weight, and they probably were now fully aquatic and unable to move on land. (See *Basilosaurus* on Plate 6.)

The development of hearing in whales is very important. The structure of the inner ear of whales differs from that of all other mammals.[38] The inner ear of whales exhibits a round thickened piece of bone called the tympanic bulla, which contains the middle ear and has a thickened medial lip called the involucrum. This is attached to a bony crest called the sigmoid process, which is S-shaped. In modern whales, this is only loosely connected and is surrounded by air pockets that help to isolate it. (See Fig. 1, The Ear in Land Mammals and Whales, p. 64.)

This unique inner ear structure allows sound to be detected underwater, mediated by a padded fatty structure under its lower jaw. Another protocetid, *Pakicetus* (discussed above), had the tympanic bulla but lacked the lower jaw fat pad of modern whales. As such, it could still detect airborne sound.[39] *Ambulocetus natans* (discussed above) also demonstrates transition to a whale anatomy of the inner ear. It is likely to have developed a fat pad under the jaw. The bone construction from the fragments remaining suggests that what sound it could detect underwater seems to have been relayed by bone conduction. Remingtonocetids and protocetids (both discussed earlier) had developed the fat pad under their jaws common to all later whale types.

38. Sirpa Nummela et al., "Sound Transmission in Archaic and Modern Whales: Anatomical Adaptations for Underwater Hearing," *The Anatomical Record* 290 (2007): 716–33.

39. Nummela, Hussain, and Thewissen, "Cranial Anatomy of Pakicetidae (Cetacea, Mamalia)," 746–59.

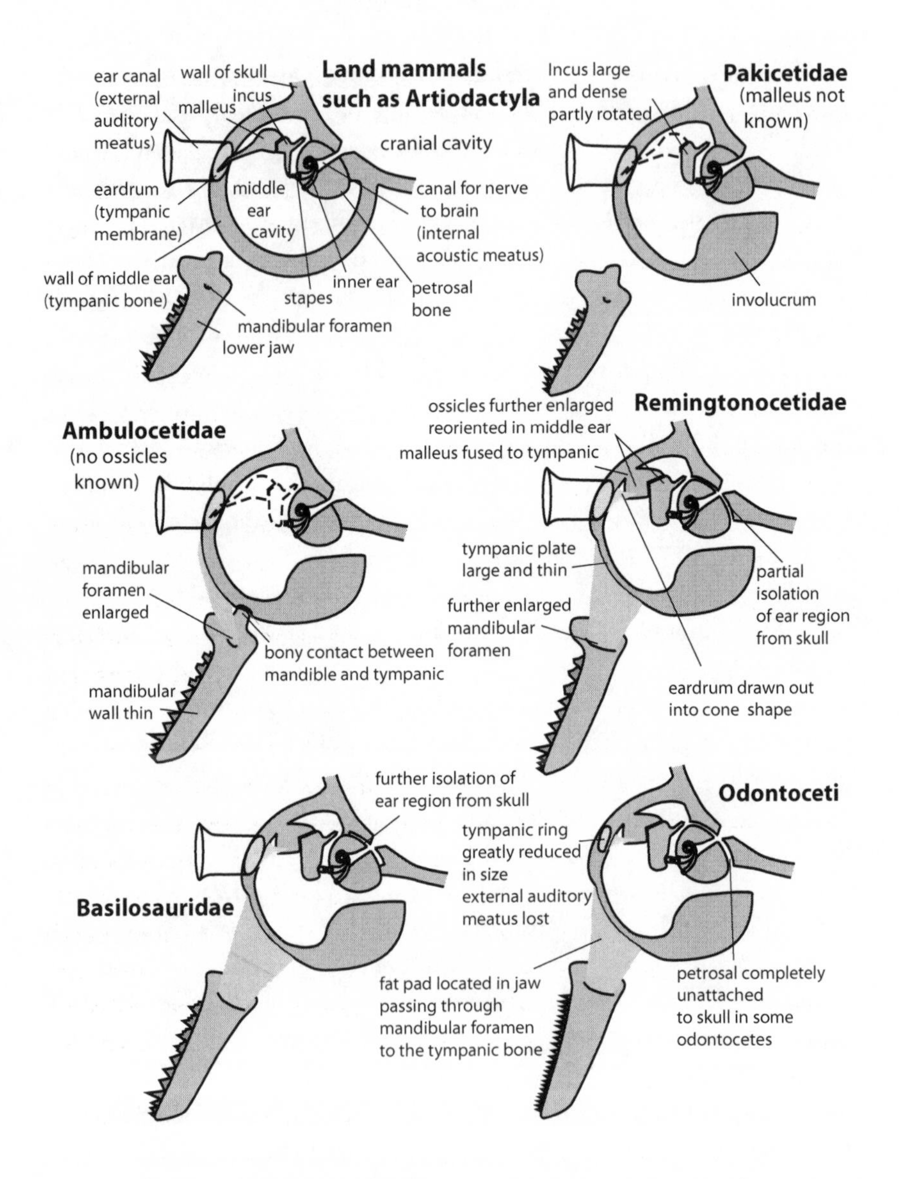

Figure 1. The Ear in Land Mammals and Whales, Figure 41 from J. G. M. "Hans" Thewissen, *The Walking Whales* © University of California Press. The diagram at the top left identifies all of the parts. Labels in other diagrams indicate which changes took place at each evolutionary step leading to modern whales. Dashed lines indicate bones not known for the group in question; their shape has been inferred from other groups.

None of these extinct whales (known under the designation Archaeoceti[40]) show evidence of the "melon" that allows directional focusing of sound. This includes *Basilosaurus* and *Dorudon*, which means that echolocation is found only in modern toothed whales. However, a newly discovered ancient porpoise named *Echovenator sandersi*, from about twenty-seven million years ago, has been shown to have been able to hear very high-frequency sounds and to have had other parts of the apparatus for echolocation.[41] This suggests that echolocation had developed among the earliest toothed whales (Odontoceti).

The evolution of cetacean nostrils is interesting to track as well. In the case of *Pakicetus,* the nostrils were on the end of its nose, as were the nostrils of *Ambulocetus natans* and the remingtonocetids such as *Kutchicetus*. However, *Rodhocetus* possessed nostrils farther back on the head. They were even farther back on *Basilosaurus* and *Dorudon*. While the blowhole on top of the head appears only in modern whales, these creatures demonstrate the intermediate forms that led to this morphological development.

Another fascinating aspect in the evolution of whales is the gradual shift in the aquatic environments that whale ancestors inhabit. A study of oxygen isotopes in cetacean remains indicates whether the creatures were taking in fresh or salt water when their teeth were formed. *Pakicetus* definitely inhabited fresh water and was most likely indigenous to lakes and streams. The fossil teeth of *Ambulocetus natans* seem to indicate that the creature lived in a range of aquatic contexts when it came to salinity. Some fossils show them inhabiting fresh water, but others indicate that they were apparently swimming in bodies of salt water. *Kutchicetus* (and other members of the family Remingtonocetidae) seems to have existed in exclusively saltwater habitats, indicating their native environments were likely coastal areas and saltwater lagoons. *Rodhocetus*, *Dorudon*, and *Basilosaurus* were demonstrably at home in a saltwater environment.

Modern experts on ancient whales are justified to assert: "whale origins form one of the most compelling examples of macroevolutionary change in vertebrates."[42] As one paleontologist surmises:

40. Term for primitive cetaceans that lived from fifty-five to twenty-three million years ago. The earliest cetaceans on record, they include the various early whales that we discussed above.

41. Morgan Churchill et al., "The Origin of High-Frequency Hearing in Whales," *Current Biology* 26.16 (22 August 2016): 2144–49.

42. Thewissen and Bajpai, "Whale Origins as a Poster Child for Macroevolution," 1037–49.

> In summarizing the paleontological evidence, we have noted the consistent changes that indicate a series of adaptations from more terrestrial to more aquatic environments as we move from the most ancestral to the most recent species. These changes affect the shape of the skull, the shape of the teeth, the position of the nostrils, the size and structure of both the forelimbs and the hind limbs, the size and shape of the tail, and the structure of the middle ear as it relates to directional hearing underwater and diving. The paleontological evidence records a history of increasing adaptation to life in the water – not just to any way of life in the water, but to life as lived by contemporary whales.[43]

Conclusions

This chapter has taken up the topic of "kinds" in Genesis and the question as to whether living things can develop beyond this conceptual boundary. The first concern is what the Bible actually says. The expression "after its kind" has been taken as an important biblical indication that there is a limit to variation in animal kinds. In recent years, creationists have even developed their own version of classification, baraminology, based on assumptions about "ark kinds," the generic animals that can develop into various species over time. The problem is that their definitions attempt to form a scientifically coherent picture. Yet, they distance themselves from the actual language of the biblical text. This problem is indeed inevitable, because the Bible is not a scientific text nor is it open to attempts to force linguistic precision when it refers to "kind" in Genesis, for the biblical language does not show that kind of precision.

In reality, there did not need to be so much disagreement over the meaning of "after its kind" in the Bible because creationists have long recognized that a certain amount of evolution – variation and change over several generations – does in fact take place. However, they have generally drawn the line at "microevolution" (small developments and changes), arguing that the biblical text denies "macroevolution" (the eventual change of living things over long periods of time to something different). Yet as we

43. Raymond Sutera, "The Origin of Whales and the Power of Independent Evidence," *Reports of the National Center for Science Education* 20.5 (2000): 33–41.

saw when carefully examining the texts, this whole position is based on an unjustified interpretation of the language of the biblical text itself. Such an interpretation is not supported by Hebrew grammar nor linguistic usage in Jewish texts nor in early translations and commentaries. Neither is this position justifiable by scientific data. There are indeed transitional forms in nature that support the understanding of paleontologists trying to reconstruct the history of evolution.

The weight of the evidence for transitional forms is so compelling that to deny them has begun to look like an exercise in self-deception. Indeed, as we have noted, some creationists give way to these interpretations while trying to reconcile them with their constructs. The problem for creationists is their understanding that the Bible itself contradicts the existence of such forms. There is no biblical evidence for the view that there is a genetic limit on how far an animal type can change. There is simply nothing in the Bible that teaches against "macroevolution." The biblical text, in fact, says nothing about evolution at all, one way or the other.

However, once someone accepts "microevolution" (as most creationists do today), there is nothing biblical or scientific that can rule out macroevolution. Young-earth creationist Kurt Wise, for example, concedes that macroevolution appears to occur in the fossil records: "Substantial supporting evidence of macroevolutionary theory can be found in the fossil record of stratomorphic intermediates."[44] He of course expresses confidence that the creationist model will find ways of developing an explanation that is superior to the evolutionary one. But the important point is that, contrary to the assertion that the theory of evolution has no evidence for macroevolution, such evidence is clearly there. Most scientists of the past two centuries have come to agree that the evidence for evolution is completely overwhelming. The Bible is not well treated when its own language is misinterpreted to oppose conclusions vastly supported by scientific study.

A number of creationist scientists have come to recognize what biologists, paleontologists, and other scientists have been describing for some time: significant changes in the morphology (shape and structure) of species do take place over time. Indeed, some creationist scientists are now arguing for an accelerated form of evolution many times faster than normally ac-

44. Wise, "Towards a Creationist Understanding of 'Transitional Forms,'" 221.

cepted by evolutionary biologists to bolster another questionable biblical conclusion about the age of the earth. Unfortunately, they arbitrarily assume that this rapid and radical change goes so far and then suddenly comes to an abrupt halt, lest an organism might change too much and jump classifications into another "kind." All this is rhetoric, without solid biblical interpretation or a measured understanding of the natural mechanisms evidenced in evolutionary history.

PART II

Evangelicals and Evolution

5
A SHIFTING CONVERSATION
Science and Religion

According to a report from the Pew Research Center ("America's Changing Religious Landscape"),[1] Christian affiliation among Americans decreased from 78.4 to 70.6 percent between 2007 and 2014. This reduction was primarily in the mainline Protestant and Catholic denominations. Non-Christian faiths were up slightly, but "unaffiliated" was up from 16.1 to 22.8 percent. Evangelical Protestants were mainly stable but still down almost 1 percent. Those reporting themselves as having no religion went up from 14.8 to 25.1 percent.

When it comes to the question of evolution, polls conducted in 2009 and 2013 surveyed the subject.[2] Overall, 60 percent of Americans accept evolution and 33 percent reject it. This is a slight decrease from 2009 when 61 percent believed in evolution. Although 57 percent of evangelical Protestants reject evolution, the majority of mainline Protestants, Catholics, and many other religious people accept it.

The Challenge of Militant Atheism

Accompanying this trend in secularization is the rise of militant atheism. One of the most outspoken of these atheists is Richard Dawkins, professor for the public understanding of science at the University of Oxford in

1. Available at http://www.pewforum.org.
2. See "Religious Landscape Study" (http://www.pewforum.org/religious-landscape-study).

England. To many scientists he is best known for his theory of the "selfish gene," a theory that pictures natural selection as taking place at the genetic level rather than merely at the level of species and adaptability. Dawkins argues that our genes are programmed to reproduce themselves and that they govern all our actions toward the goal of maximizing the chance to pass on genes to the new generation. This idea rules out certain spiritual and ethical concepts like altruistic behavior. Some scientists do not accept the theory, while a number of those who accept the idea of natural selection at the level of genetics do not accept that this does away with selflessness.

Dawkins is widely known for his book, *The God Delusion*.[3] He writes with great emotion that we should be rational and make rational decisions. His contempt for religion knows no bounds. For some reason, he sees no conflict between passionate atheism, with boundless antipathy toward religion, and his stance of rationality. The irony of what he asserts seems entirely to have escaped him. What is grating about Dawkins is not his atheism: he is welcome to believe – or not believe – what he wishes. One big problem is his stance that all should be rational when it is clear that his hatred of religion is in part irrational.

Another problem is his view – which he shares surprisingly enough with some religious people – that belief in evolution is incompatible with a belief in God and a religious faith. Dawkins is in fact a man of great faith! He believes earnestly, one might say *religiously*, that science will ultimately provide all the answers about who we are and where we came from. This belief is simply not shared by many scientists (or philosophers, for that matter). Most scientists recognize the limits of science to explain reality.

Dawkins claims that people are becoming better – more considerate of others, more tolerant, and less violent – as religion declines. The question of whether the world is getting better could be debated at length. Dawkins's belief that improvements in the way people treat each other correspond with the decline of religion is a logical error and is an assertion of blind faith. The decline of religion is also not a universal. Historians who have examined many of the wars in the twentieth century – the most destructive in the history of our world – would not describe them for the most part as religious wars.[4]

3. Richard Dawkins, *The God Delusion* (London: Bantam, 2006).

4. This is ably argued by William T. Cavanaugh, *The Myth of Religious Violence: Secular Ideology*

Yet, the pronouncements of Dawkins and his fellows are not the full story, not by any means. True, the public manifestation of militant atheism ultimately has an effect on many people. Many readers will have the impression that scientists are basically against religion. They will have seen the newspaper headlines that this scientist has attacked religion as nothing but superstition or that scientist has expressed the confident opinion that religion is shrinking in the modern world and will gradually disappear. In actuality, the past few years have muted the voice of those who thought religion would disappear. Religion has been on the rise in China. Islam continues to thrive. In fact, a number of scientists are willing to accept the place of religion.

Science versus Religion?

What percentage of scientists express some sort of faith perspective? It is difficult to quantify the beliefs of scientists. One survey of one thousand American scientists from a variety of fields found that 40 percent of them believed in a personal God.[5] However, if one looks more closely at the data, it seems that certain specific types of scientists were more inclined to reject any notion of personal faith. For example, in a survey of 255 biological and physical scientists who were members of the US National Academy of Sciences, only 7 percent professed belief in a personal God.[6] Another survey – this time of 151 evolutionary biologists who were members of the various national academies of science in twenty-one countries – found that only 5.4 percent believed in a personal God.[7] Many distinguished scientists are in-

and the Roots of Modern Conflict (Oxford: Oxford University Press, 2009). Cavanaugh accepts that some religions or aspects of religion do indeed promote violence, but this is not inevitable: there is no essential aspect of religion that makes it more inclined to violence than any "secular" ideology. His argument is that the stereotyped conclusion of religion causing violence is not an empirically demonstrable fact but an ideology due to shifts in power and authority in the development of the modern Western world.

5. Edward J. Larson and Larry Witham, "Scientists Are Still Keeping the Faith," *Nature* 386 (1997): 435–36.

6. Edward J. Larson and Larry Witham, "Leading Scientists Still Reject God," *Nature* 394 (1998): 313.

7. Gregory W. Graffin and William B. Provine, "Evolution, Religion and Free Will," *American Scientist* 95.4 (2007): 294–97.

deed atheists or at least agnostics. However, it is another matter among the rank and file in the scientific community. There the percentage who reject a personal deity is much smaller. If one takes the scientific community as a whole, a significant proportion of practicing scientists do believe in some version of God.[8]

Views of Some Nonreligious Scientists

Despite the numbers above, there are many scientists who have a personal religious belief but see no conflict between their science and their religious faith. It is also reassuring that a number of well-known scientists who do not have a faith perspective nevertheless argue that science and religion are not in opposition to each other. Indeed, one survey — already noted above — found that only 10 percent would agree with Professor Dawkins that religion and evolutionary science were in fundamental conflict. On the contrary, nearly 85 percent thought that there was no inevitable conflict.

The late but famous evolutionary biologist George Gaylord Simpson, a scientist not speaking from the vantage point of any faith, once wrote:

> There is neither need nor excuse for postulation of non-material intervention in the origin of life, the rise of man, or any other part of the long history of the material cosmos. Yet the origin of that cosmos and the causal principles of its history remain unexplained and inaccessible to science. Here is hidden the First Cause sought by theology and philosophy. The First Cause is not known, and I suspect it will never be known to living man. We may, if we are so inclined, worship it in our own ways, but we certainly do not comprehend it.[9]

The late Stephen J. Gould, who openly identified himself as an agnostic, strongly argued in favor of what he called *NOMA*: "nonoverlapping magis-

8. Warren D. Allmon, "The 'God Spectrum' and the Uneven Search for a Consistent View of the Natural World," in *For the Rock Record: Geologists on Intelligent Design*, ed. Jill S. Schneiderman and Warren D. Allmon (Berkeley: University of California Press, 2009), 193.

9. George Gaylord Simpson, *The Meaning of Evolution*, rev. ed. (New Haven: Yale University Press, 1967), 279.

teria" of religion and science.[10] This idea would allow science to speak of the description of the natural world and its phenomena, while religion would speak to the moral world. Science and religion were concerned with separate spheres of influence, study, and inquiry. This, he felt, would allow the two ways of looking at the world to coexist peacefully. While his perspective on religion was criticized by some reviewers as reducing it to morality,[11] his instincts do provide one way forward.

Niles Eldredge, a close associate of Gould, identified himself as a "lapsed Baptist." He writes: "Along with many others, I see myself as an agnostic because 'atheist' is too definitive, implying one can know something that is in principle unknowable."[12] Yet Eldredge is not hostile to the beliefs of others: "I think that concepts of God – *all* concepts of God – are about something, and of course I am not about to quarrel with anyone's personal interpretation of any one of those particular concepts of God."[13] When he saw the Edward J. Larson and Larry Witham poll that 40 percent of the scientists surveyed had expressed a belief in a personal God, he reflected:

> I was not surprised by their results. I myself may not affect such belief, but I know a number of colleagues who do. The number of religious scientists grows, of course, when one expands belief to a less proactive conception of a Christian God, or, of course, acknowledges that other religions – Judaism, Islam, Buddhism, Hinduism, and so on – are, well, actual religions, equally valid per se as the narrow-minded version of Christianity espoused by many, if not all, creationists.[14]

Eldredge's acceptance of the faith of scientists, whatever it might be, does not extend to what he understands as the "cultural war" – a political

10. This is discussed most fully in Stephen J. Gould, *Rock of Ages: Science and Religion in the Fullness of Life*, Library of Contemporary Thought (New York: Ballantine Publishing, 1999).

11. See the discussion and references in Allmon, "The 'God Spectrum' and the Uneven Search for a Consistent View of the Natural World," 187–89.

12. Niles Eldredge, *The Triumph of Evolution and the Failure of Creationism* (New York: W. H. Freeman, 2000).

13. Eldredge, *The Triumph of Evolution and the Failure of Creationism*, 17.

14. Eldredge, *The Triumph of Evolution and the Failure of Creationism*, 169.

battle — that is being instigated by the demands of creationism about such things as teaching science, including evolution, in schools.

Michael Ruse is a well-known philosopher of science with an interest in the relationship between religion and science. He has written a number of books on the subject, including *Can a Darwinian Be a Christian? The Relationship between Science and Religion* and *The Evolution-Creation Struggle.*[15] Although Ruse was raised a Quaker, he lost his faith as a young adult and currently thinks of himself as "an agnostic on deities and ultimate meanings," though he prefers the designation of "skeptic."

Eugenie C. Scott has been executive director of the National Center for Science Education and also president of the American Association of Physical Anthropologists. She describes her background as liberal Protestant. At present, she is a secular humanist and atheist. Yet, Scott does not belittle spirituality: "Science is a limited way of knowing, looking at just the natural world and natural causes. There are a lot of ways human beings understand the universe — through literature, theology, aesthetics, art or music."[16]

Phil Senter, a biologist at Fayette State University in North Carolina, grew up as a creationist until studying evolution in high school. He is noted for applying the techniques of creation science in his research to test out creationist claims on their own terms. Although most biologists do not accept the particular application of this technology in the way worked out by baraminologists, Senter decided to apply these techniques just as the authors themselves did. He concluded that the arguments support evolution![17]

Paleontologist Donald Prothero, who has made his name in the study of mammalian evolution, has written many books, both scholarly and popular. One of his most influential books is his recent *Evolution: What the Fossils Say and Why It Matters.* He writes the following in a section entitled "To the Reader: Is Evolution a Threat to Your Religious Beliefs?":

15. Michael Ruse, *Can a Darwinian Be a Christian? The Relationship between Science and Religion* (Cambridge: Cambridge University Press, 2000). See also Michael Ruse, *The Evolution-Creation Struggle* (Cambridge: Harvard University Press, 2005).

16. Monica Lam, "Profile/Eugenie Scott/Berkeley Scientist Leads Fight to Stop Teaching of Creationism," *San Francisco Chronicle*, February 7, 2003.

17. Senter, "Using Creation Science to Demonstrate Evolution 1," and Senter, "Using Creation Science to Demonstrate Evolution 2."

> Many people find the topic of evolution and religion troubling and confusing: Some were raised in very strict churches that preached that evolution is atheistic and that to even think about the evidence of evolution is sinful. Fundamentalists have long tried to drive a wedge between traditional Christians and science, arguing that their interpretation of the Bible is the only one and that anyone who accepts the evidence for evolution is an atheist.[18]

Prothero refutes this position as a complete misunderstanding:

> But this not true. The Catholic Church, along with most mainstream Protestant and Jewish denominations, has long ago come to terms with evolution and accepted it as the mechanism by which God created the Universe. The Clergy Letter Project includes the signatures of more than 10,000 ministers, priests, and rabbis in the United States who accept evolution and do not view it as incompatible with religious belief.[19]

In a recent news story, the theoretical physicist Peter Higgs spoke to the question of science and religion.[20] Although Higgs says that he is not himself a religious believer, in an interview with the Spanish paper *El Mundo*, he had some outspoken comments about the approach of Richard Dawkins: "Dawkins in a way is almost a fundamentalist himself, of another kind." He found Dawkins's approach to religion embarrassing. Instead, Higgs argues for a more tolerant approach: "The growth of our understanding of the world through science weakens some of the motivation which makes people believers. But that is not the same thing as saying they are incompatible. . . . Anybody who is a convinced but not a dogmatic believer can continue to hold his belief. It means I think you have to be rather more careful about the whole debate between science and religion than some people have been in the past."[21]

18. Donald R. Prothero, *Evolution: What the Fossils Say and Why It Matters* (New York: Columbia University Press, 2007).

19. Prothero, *Evolution*, xvii. See the clergy letter website: www.theclergyletterproject.org.

20. The one who coined the term "Higgs Boson," a term that refers to an elementary particle that should allow physicists to validate the last untested area of the Standard Model's approach to fundamental particles and forces.

21. Quotations from Alok Jha, "Peter Higgs Criticises Richard Dawkins over Anti-religious 'Fundamentalism,'" *The Guardian*, December 26, 2012. The original interview appeared in Pablo

The Astronomer Royal (a crown appointment) and also head of the Royal [Scientific] Society is Lord Martin Rees. In a recent interview, he said that he believed science and religion belonged to different domains and could coexist peacefully. He also noted that theologians who take their theology seriously do not attempt to use theology to explain the mysteries of science. His acceptance of the coexistence of science and religion has not endeared him to all his fellow scientists (Dawkins called him a "compliant Quisling"), but he told the interviewer, "I've got no religious beliefs at all."

This is only a small sample of scientists expressing support for the view that science and religion are potentially compatible. Many scientists and other scholars argue that "science versus religion" is a false dichotomy. For most scientists without a faith perspective, there is no necessary conflict between science and religion, though most would prefer that each keeps to its own domain. Most would accept that scientists have no more right to pronounce dogmatically on religion than theologians or people of faith have a right to reject the claims of science.

The comments of some of these individuals have introduced a more specific question: Can someone remain a believing Christian and also accept the scientific theory of evolution? Likewise, can someone be a practicing Jew or Muslim and reconcile these faith perspectives with the theory of evolution? For Dawkins, there is no doubt that this is impossible. Indeed, for one who is a scientist, he is remarkably dogmatic about this! The fact is that Dawkins's approach puts him in a rather ironic position. By stating that acceptance of the science of evolution requires one to be an atheist, he is, in fact, holding a point of view of one of his intellectual foes. Some creationists would argue the very same thing.

Conclusion

We have examined the assertions of the well-known scientist and atheist Richard Dawkins that science and religion are incompatible. We have also surveyed the views of scientists who claim there is no such inherent con-

Jáuregui, "Peter Higgs: 'No soy creyente, pero la ciencia y la religión pueden ser compatibles,'" *El Mundo*, December 27, 2012.

flict. It is clear that the conflict between science and religion – indeed, between religion and evolution – that Dawkins espouses is rejected by many scientists. Many scientists on both sides of the personal faith divide have expressed the view that science and religion can coexist in harmony and integrity, especially as long as each does not speak into the domain prescribed by the other.

Well-respected science writer Colin Tudge in his recent book, *Why Genes Are Not Selfish and People Are Nice*, has gone further, however, and argued with vigor that "science is limited. It does not tell us all there is to know, or what we might reasonably want to know."[22] Tudge is by no means castigating science: "We can accept that science is wonderful, and that rationality is vital – but then acknowledge too that if we are truly to get to the truth of things (or at least to get as close as possible) then we must venture beyond science, and beyond what is conventionally called rationality."[23] His solution – as a scientist but also as a humanist – might be shocking to some: "A worldview is possible and necessary that embraces both science and the core ideas of religion. . . . A worldview that truly embraced them both would, I suggest, take us as close to truth as it is possible for human beings to approach."[24]

Today, many believing scientists do embrace evolution as the best scientific explanation for how life arose, as we have already seen. But what do religion experts (theologians and the like) think about evolution? Is there a necessary conflict between being a person of faith and believing in evolution? Let us now look at some key figures from the other side of the equation to see what theologians and religious thinkers have made of this tension.

22. Colin Tudge, *Why Genes Are Not Selfish and People Are Nice: A Challenge to the Dangerous Ideas that Dominate Our Lives* (Edinburgh: Floris Books, 2013), 223.

23. Tudge, *Why Genes Are Not Selfish and People Are Nice*, 224.

24. Tudge, *Why Genes Are Not Selfish and People Are Nice*, 240.

6

SCIENCE AND FAITH

The Believing Scientist

In the previous chapter, we explored the range of views about religion among scientists, focusing on authors who write on evolution. As we saw, there are a range of views: a few are hostile toward any form of religion and make a point of proclaiming this publicly, but many others see no incompatibility between science and religion, even though they may themselves not be religious. In the present chapter, we want to look at those paleontologists and evolutionary biologists who are themselves people of faith (whether Christian, Jew, Muslim, or otherwise).

Now we come to a critical question: Can someone be a scientist and also have a religious belief? In fact, many scientists, including biologists and paleontologists, are quite happy to profess a personal life of faith. Such scientists see no conflict between their religious beliefs and their work in science. Let us consider the voices of believing scientists.

Views of Believing Scientists

Robert T. Bakker, a paleontologist and curator of paleontology for the Houston Museum of Natural Science, is particularly known for expounding the idea of his teacher John Ostrom that at least some dinosaurs were warm blooded (endothermic).[1] Bakker also came up with the thesis that

1. See his book, Robert T. Bakker, *The Dinosaur Heresies: New Theories Unlocking the Mystery of the Dinosaurs and Their Extinction* (New York: William Morrow, 1986).

some dinosaurs had feathers. This has now been demonstrated by ample fossil finds and is accepted by almost all paleontologists. However, Bakker is also a Christian, even an ordained Christian minister. He argues that there is no real conflict between religion and science. What is interesting is that this man of science quotes freely from one of the church fathers, Augustine, who argued against a literal understanding of the book of Genesis, seeing the meaning of the text as spiritual rather than literal.

R. J. Berry, professor of genetics at University College London until his retirement, has published several books that include contributions by scientists with religious beliefs. In *Real Science, Real Faith*, for example, sixteen scientists openly discuss their faith.[2] More recently he followed that up with *Real Scientists, Real Faith*, in which eighteen top scientists with a firm Christian faith tell their story.[3]

Some scientists are people of faith (Christians, Jews, Muslims, or other) and yet do not at all find this to be incompatible with their scientific understanding. Perhaps one of the best known is the Cambridge paleobiologist Simon Conway Morris. He first made his name as a graduate student by his work on the Burgess Shale. This is one of the world's most impressive fossil fields, famous for the way the soft parts of extinct organisms are exceptionally well preserved.[4] Morris describes the importance of the work of Christian thinkers such as G. K. Chesterton and even Dorothy L. Sayers in being formative in his spiritual journey. He is noted among other things for his work on the "Cambrian explosion" – which refers to the fairly rapid appearance, around 542 million years ago, of most major animal phyla in the fossil record. He not only makes use of evolutionary theory but also provides important confirmation of the essential correctness of evolution.

Two other renowned Cambridge scientists grew up in a Christian context and retained their Christianity as they embraced a life of science. Denis R. Alexander has served for years at the Babraham Institute in Cambridge and has also been chairman of the Molecular Immunology Depart-

2. R. J. Berry, ed., *Real Science, Real Faith* (Crowborough: Monarch Publications, 1991).

3. R. J. Berry, ed., *Real Scientists, Real Faith* (Oxford: Monarch Books, 2009).

4. See Stephen J. Gould, *Wonderful Life: The Burgess Shale and the Nature of History* (London: Hutchinson Radius, 1989). Also, Simon Conway Morris, *The Crucible of Creation: The Burgess Shale and the Rise of Animals* (Oxford: Oxford University Press, 1998).

ment, after teaching many years at several universities in the Middle East. He has written several books on the relationship of science and religion. One of his first books, *Beyond Science*, was an offering on this subject.[5] He has also written *Rebuilding the Matrix – Science and Religion in the 21st Century*.[6] He discusses DNA, natural selection, Genesis, Adam and Eve, the fall, and intelligent design. He writes as one who believes "that the Bible is the inspired Word of God from cover to cover."[7] Yet he concludes that personal "saving faith through Christ in the God who has brought all things into being and continues to sustain them by his powerful Word, is entirely compatible with the Darwinian theory of evolution."[8] Thus, his answer to the question in his title – Do we have to choose between creation and evolution? – is negative. Robert White is a professor of geophysics at Cambridge. He and Denis Alexander collaborated in founding the Faraday Institute for Science and Religion in 2006 at St. Edmund's College, Cambridge. The institute works to provide a context for interdisciplinary academic research at the intersection of science and religion.

Francis Collins earned a PhD in chemistry and then became a physician. He is noted as the director of the Human Genome Research Institute at the National Institutes of Health in Bethesda, Maryland, for the years 1993 to 2008. He was an agnostic who inclined toward atheism. His academic career progressed well. Something changed when he was challenged by a terminally ill patient as to his beliefs. That encounter prompted him to critically examine the question of God's existence. In time, he converted to Christianity. He chronicles the story of his coming to faith in Jesus in his book, *The Language of God*, which also describes his respected work in DNA.[9]

John Polkinghorne initially completed a PhD in physics and then carried out postdoctoral work at the California Institute of Technology.[10]

5. Denis Alexander, *Beyond Science* (Philadelphia: A. J. Holman, 1972).

6. Denis Alexander, *Rebuilding the Matrix: Science and Faith in the 21st Century* (Grand Rapids: Zondervan, 2003). See also Denis Alexander, *Creation or Evolution: Do We Have to Choose?* 2nd ed. (Oxford: Monarch Books, 2014).

7. Alexander, *Creation or Evolution: Do We Have to Choose?*, 11.

8. Alexander, *Creation or Evolution: Do We Have to Choose?*, 351.

9. Francis S. Collins, *The Language of God: A Scientist Presents Evidence for Belief* (London: Pocket Books, 2007).

10. The information on Polkinghorne comes primarily from the Introduction to Thomas Jay Oord, ed., *Polkinghorne Reader: Science, Faith and the Search for Meaning* (London: SPCK,

While still a student he felt a call for a deeper commitment to his Christian faith. He was appointed professor of mathematical physics at Cambridge in 1968. In 1979 he resigned his professorship and undertook training to be an Anglican priest, being ordained in 1982 and doing parish work for the next few years. However, he was invited to become dean and chaplain of Trinity Hall, Cambridge, in 1986, and shortly afterward was appointed president of Queen's College, Cambridge. He was knighted in 1997. He has written extensively on science and religion, using the analogy that science and religion are like two separate eyes of a pair of binoculars that enables us to see more clearly. Fully accepting evolution, he believes the universe is billions of years old, as do other scientists.

Alister E. McGrath was an atheist when he went to Oxford to study chemistry and biochemistry. He also studied the philosophy of science, which led him to explore the limits of science. McGrath began to realize that his atheism was largely unexamined. After sustained reflection and careful scrutiny, he abandoned atheism. In 1977 he completed his doctorate in molecular biophysics. Because of his new faith, McGrath began to study theology, eventually earning a doctor of divinity degree and becoming a professor in historical theology at Oxford. He is currently Andreas Idreos Professor in Science and Religion at the University of Oxford and is professor of divinity at Gresham College; earlier he had been professor of theology, ministry, and education at Kings College London and head of the Centre for Theology, Religion and Culture. He remains critically engaged with the questions of science and faith and has written, among many other books, a three-volume *Scientific Theology* and *The Dawkins Delusion? Atheist Fundamentalism and the Denial of the Divine*.[11]

Patricia H. Kelley is an earth scientist at the University of North Carolina at Wilmington and has been president of the Paleontological Society. She is married to an ordained Presbyterian minister. She has written and lectured on why she believes that there is no conflict between her faith perspective and her science profession. In one article, "Teaching Evolution during the Week and

2010). Polkinghorne has also written an autobiography: John Polkinghorne, *From Physicist to Priest: An Autobiography* (London: SPCK, 2007).

11. Alister E. McGrath, *A Scientific Theology*, 3 vols. (London: T&T Clark, 2001–2003); Alister E. McGrath and Joanna Collicutt McGrath, *The Dawkins Delusion? Atheist Fundamentalism and the Denial of the Divine* (Downers Grove, IL: InterVarsity Press, 2010).

Bible Study on Sunday,"[12] she describes how she both teaches evolution in her university classes and the Bible to an adult Bible study class in church. Although she believes the Bible is the word of God and authoritative in her life, she is not a biblical literalist. She reflects: "Science and religion are different ways of knowing. . . . Religion provides answers to questions of ultimate meaning."[13]

Kenneth R. Miller, a biologist at Brown University, approaches questions of origins from a scientific point of view. In his book, *Finding Darwin's God: A Scientist's Search for Common Ground between God and Evolution*, he asks readers to consider afresh the ultimate question: "Where are you from?" His answer is an interesting mix of hard science, including Darwinism. Yet, Miller expresses his own faith perspective as well. He concludes that evolution and a religious faith need not be in conflict: "If the Creator uses physics and chemistry to *run* the universe of life, why would not He have used physics and chemistry to *produce* it, too?"[14]

Robert J. Asher is a vertebrate paleontologist at Cambridge University and also curator of vertebrates in the University Museum of Zoology. The title of his recent book, *Evolution and Belief: Confessions of a Religious Paleontologist*, reveals his understanding of the link between religious faith and evolution. Much of the book discusses science, drawing on his research and expert knowledge on mammals and other vertebrates. Yet he also carefully considers the relationship of science and religion. His "confession" of the title is "the fact that I do not see a contradiction between my profession as an evolutionary scientist and my belief in God."[15]

Joan Roughgarden employs a similar subtitle for her book, *Evolution and Christian Faith: Reflections of an Evolutionary Biologist*. She is professor of biological sciences and of geophysics at Stanford University. Roughgarden is also active in the Episcopal (Anglican) Church and the daughter of Episcopal missionaries. She drifted away from her faith after going to university but later returned to think through the relationship of her faith and her

12. Patricia H. Kelley, "Teaching Evolution during the Week and Bible Study on Sunday," in *For the Rock Record: Geologists on Intelligent Design*, ed. Jill S. Schneiderman and Warren D. Allmon (Berkeley: University of California Press, 2009), 163–79.

13. Kelley, "Teaching Evolution during the Week and Bible Study on Sunday," 168.

14. Kenneth R. Miller, *Finding Darwin's God: A Scientist's Search for Common Ground between God and Evolution* (New York: Harper Perennial, 1999), 268.

15. Asher, *Evolution and Belief*, xiii–xiv.

occupation as an evolutionary biologist. She uses a creationist term to describe her point of view as "theistic evolution." She explains: "God creates the world according to a plan that continues to unfold, in the fullness of time, through the natural processes that science studies."[16]

Other scientists could be mentioned. Francisco J. Ayala was professor of biological sciences at the University of California and a Christian and former Dominican priest. He wrote a little book, *Darwin and Intelligent Design*, in which he argued that science (including evolution) was not incompatible with religion.[17] Darrell R. Falk was professor of biology at Point Loma Nazarene University. He discovered as a university student that biology did not need to be alien territory for the Christian. Falk relates his experience in *Coming to Peace with Science: Bridging the Worlds between Faith and Biology*.[18] As both an evangelical Christian and a scientist, he accepts an ancient earth and an evolved creation.

Geologists with a Faith Perspective

It is not just biologists and paleontologists who must deal with the question of evolution. Major aspects of evolution depend on questions of dating, the age of the earth, and the geology under our feet. Some believers have the impression that most geologists of faith interpret the history of the earth in terms of the biblical flood and reject an ancient earth. In fact, almost all geologists of a religious faith nevertheless see the geology in the rocks with a scientific eye that does not differ in its essentials from that of other geologists. Far from seeing a young earth, they date the rocks in the same way as their colleagues who have no personal religious belief.

Unfortunately, an untrained geologist in the early part of the twentieth century had a great influence on evangelical Christianity by interpreting the earthly strata according to his view of the flood.[19] George

16. Joan Roughgarden, *Evolution and Christian Faith: Reflections of an Evolutionary Biologist* (Washington, DC: Island Press, 2006), 34.

17. Francisco J. Ayala, *Darwin and Intelligent Design* (Minneapolis: Fortress Press, 2006).

18. Darrell R. Falk, *Coming to Peace with Science: Bridging the Worlds between Faith and Biology* (Downers Grove, IL: InterVarsity Press, 2004).

19. For information on George McCrady Price and other figures in the creationist move-

McCrady Price (1870–1963), a young Seventh Day Adventist who began as a construction worker and received no formal training in geology, attempted to overturn conventional geology. His ideas were picked up by certain evangelical writers and made popular. Yet many of those who have read or heard of these ideas are not aware that Price's main student and later associate, Frank Lewis Marsh, later abandoned Price's ideas. This was because of Marsh's own experience in the field, in which his empirical observations failed to uphold the views of his mentor. Marsh defended one of the main geological backbones that Price had seriously questioned: the geological column.

Gregg R. Davidson is professor and chair of the Department of Geology and Geological Engineering at the University of Mississippi. He wrote the book *When Faith and Science Collide: A Biblical Approach to Evaluating Evolution and the Age of the Earth*. In it he found evolution and the age of the earth compatible with his faith. As both a geologist and a Bible-believing Christian, Davidson points out, "much of the geologic column was roughed out long before Darwin's theory of evolution. The sequence of time periods designated as Cambrian, Silurian, Devonian, Carboniferous, and Permian had been worked out from outcrops in England, Wales, and Russia nearly 20 years before the 1859 publication of the *Origin of Species*."[20]

Davis Young and Ralph Stearley, Christian geologists, have written a book summarizing their views on geology.[21] They reject the concept of a literal worldwide flood that many read into Genesis 6–9. They consider a number of examples that allegedly support the Noachian deluge but conclude: "Although all these examples are fascinating, they do not demand catastrophic deposition of thousands of feet of sediment over the span of a year. . . . These catastrophes are observable processes. Storms, earthquakes, tsunamis, volcanic eruptions, floods, mudslides and the like are all exam-

ment, see Ronald L. Numbers, *The Creationists: From Scientific Creationism to Intelligent Design*, rev. ed. (Cambridge: Harvard University Press, 2006).

20. Gregg R. Davidson, *When Faith and Science Collide: A Biblical Approach to Evaluating Evolution and the Age of the Earth* (Oxford, MS: Malius Press, 2009), 182–83.

21. Davis A. Young and Ralph F. Stearley, *The Bible, Rocks, and Time: Geological Evidence for the Age of the Earth* (Downers Grove, IL: InterVarsity Press, 2008).

ples of brief, local catastrophes that may be responsible for much fossil preservation."[22]

In other words, a global flood would not have laid down such steady sediments. The mass graveyards are not best explained by one worldwide catastrophic event. Rather, series of local catastrophes buried and fossilized local fauna and flora.

Roger G. Wiens holds a doctorate in physics and geology and is currently a staff scientist in the Space and Atmospheric Sciences group at Los Alamos National Laboratory. Wiens has written a useful succinct guide called "Radiometric Dating: A Christian Perspective."[23] This summary of thirty-five pages provides a thorough and clear introduction to the principles of radiometric dating. It also discusses the reliability of the method in detail. He shows that radiometric methods can be tested against one another. In addition, they can be checked against other dating methods, such as tree-ring dating, ice cores, varves (an annual layer of sediment or sedimentary rock) and other annual-layer measurements, thermoluminescence,[24] electron spin resonance,[25] and cosmic-ray exposure dating. Wiens summarizes: "All of the different dating methods agree – they agree a great majority of the time over millions of years of time. Some Christians make it sound like there is a lot of disagreement, but this is not the case."[26] He concludes: "Many Christians have been led to distrust radiometric dating and are completely unaware of the great number of laboratory measurements that have shown these methods to be consistent. Many are also unaware that Bible-believing Christians are among those actively involved in radiometric dating."

22. Young and Stearley, *The Bible, Rocks, and Time*, 279.

23. Roger C. Wiens, "Radiometric Dating: A Christian Perspective" (www.asa3.org/ASA/resources/Wiens.html).

24. Thermoluminescence dating measures the accumulated radiation of the time elapsed since material containing crystalline minerals was either heated (lava, ceramics) or exposed to sunlight (sediments).

25. A technique for studying materials with unpaired electrons.

26. Wiens, "Radiometric Dating: A Christian Perspective." See also Davidson, *When Faith and Science Collide*, 119: "For radioactive methods to be grossly in error as young-earth advocates claim, it requires that two completely unrelated processes (tectonic plate movement and radioactive decay) dramatically slowed down over time at exactly the same rate so that measurements today only *coincidentally* appear to confirm the accuracy of radioactive dating methods."

It is worth taking the time to tell the story of Glenn R. Morton. He was converted to Christianity and accepted young-earth creationism while still in school. When he began his work as a field geologist, he fervently defended the dogma of a young earth. However, Morton soon discovered that careful analysis did not support the flood geology that he had been taught:

> I would see extremely thick (30,000 feet) sedimentary layers. One could follow these beds from the surface down to those depths where they were covered by vast thicknesses of sediment. I would see buried mountains which had experienced thousands of feet of erosion, which required time. Yet the sediments to those mountains had to have been deposited by the flood, if it [young-earth creationism] was true.

He went on to describe how he tried to deal with the conflict:

> I worked hard over the next few years to solve these problems. . . . I would listen to ICR [the Institute for Creation Research], have discussions with [ICR creationist] people. . . . In order to get closer to the data and know it better, with the hope of finding a solution, I changed subdivisions of my work in 1980. I left seismic processing and went into seismic interpretation where I would have to deal with more geological data.

What was the result? Did he reconcile the actual data with flood geology theory?

> My horror at what I was seeing only increased. There was a major problem; the data I was seeing at work, was not agreeing with what I had been taught as a Christian. Doubts about what I was writing and teaching began to grow. Unfortunately, my fellow young earth creationists were not willing to listen to the problems. No one could give me a model which allowed me to unite into one cloth what I believed on Sunday and what I was forced to believe by the data Monday through Friday.[27]

27. Glenn R. Morton, "The Transformation of a Young-Earth Creationist," *Perspectives on Science and Christian Faith* 52 (June 2000): 81–83.

Morton almost abandoned Christianity, but he eventually came to the conclusion that the problem was his understanding of Genesis and the way that young-earth creationists distorted the geological data to fit their particular interpretation of Genesis. He queried some of the graduates from the Institute for Creation Research, graduates he had helped to find jobs in the petroleum industry. He asked: "From your oil industry experience, did any fact that you were taught at ICR, which challenged current geological thinking, turn out in the long run to be true?" Morton himself could not think of a single axiom of young-earth geological dogma that was supported by his study of the rocks and neither could any of the ICR graduates he questioned.

Ken Wolgemuth, Gregory S. Bennett, and Gregg Davidson, three Christian geologists, made this very case in a lecture given at the Evangelical Theological Society. They challenged the "unnecessary" dilemma often presented in the writings of many young-earth creationists, namely that it was a choice of accepting either the standard interpretation of geology or the correctness of the biblical account. They, along with a growing number of other Christians working in the field of geology, utterly reject this "false dilemma." Wolgemuth, Bennett, and Davidson made it clear that they "fully believe the biblical accounts of Creation and the Flood." Yet they affirm: "It is our conviction that this position [flood geology] is unreasonable from both a biblical and scientific perspective." They further conclude:

> But could a tremendously violent flood account for the myriad layers of the earth's rocks and sediments, as well as most fossils? Flood Geology advocates would have us believe there is evidence on both sides of this question that must be weighed. Our observation is that honestly presented evidence leaves nothing left to debate. Deposition of all the earth's layers by a single flood is not only implausible, but utterly impossible unless God temporarily suspended His natural laws in order to establish layers and fossil beds that would subsequently communicate a story vastly different than what actually happened.[28]

28. Ken Wolgemuth, Gregory S. Bennett, and Gregg Davidson, "Theologians Need to Hear from Christian Geologists about Noah's Flood" (lecture, Evangelical Theological Society, November 18, 2009).

This is the overwhelming experience of geologists with a faith perspective as we have seen. The so-called "flood geology" cannot be reconciled with the actual geology in the ground: the rocks simply tell another story.

Many believing scientists do embrace evolution as the best scientific explanation for how life arose, as we have already seen. But what do theologians and biblical scholars think about evolution? Is there a necessary conflict between being a person of faith and believing in evolution?

What Do Theologians and Biblical Scholars Say?

Many people assume that all believers accept a literal interpretation of creation in Genesis 1. Many Christians and other peoples of faith think that those who trust the Bible must accept a creation week of six literal days and an earth that is only a few thousand years old. This is, of course, the impression that Richard Dawkins wishes to create: all religious people are "bigoted" and against science.

In fact, if we survey the views of Christians through the ages, we find that many – including famous past theologians – believe that Genesis does not contradict the findings of physics, geology, and paleontology about the age of the earth or development of life from simpler forms. The fact is that many biblical scholars and theologians see no conflict between the Bible and evolution.[29]

In the history of philosophy, the question of God has often preoccupied thinkers. Thomas V. Morris edited an important volume, *God and the Philosophers: The Reconciliation of Faith and Reason*.[30] Here, twenty philosophers (including himself) offer short essays describing their personal faith and beliefs. A philosopher at the University of Notre Dame for fifteen years, Morris has made important contributions to the philosophy of religion. In this book, he and his colleagues assert that faith and reason are not in-

29. For a good survey of what different groups and individuals have thought about evolution and, especially, about the age of the earth, see the historical survey throughout Davis A. Young, *The Biblical Flood: A Case Study of the Church's Response to Extrabiblical Evidence* (Grand Rapids: Eerdmans, 1995).

30. Thomas V. Morris, *God and the Philosophers: The Reconciliation of Faith and Reason* (Oxford: Oxford University Press, 1994).

compatible. Similarly, Malcolm A. Jeeves and R. J. Berry examine questions of evolution and creation in *Science, Life, and Christian Belief: A Survey of Contemporary Issues*.[31] They find that evolution and creation, Genesis and Darwin, faith and reason are not incompatible.

Pierre Teilhard de Chardin, a French Catholic theologian and geo-paleontologist (1881–1955), attempted to reconcile his mystical theology with Darwinian evolution.[32] Two of his widely read books, *The Divine Milieu* and *The Human Phenomenon*, led to a series of formal warnings from the Holy Office to bishops and heads of seminaries that his writings contained fundamental doctrinal errors and to a prohibition of Teilhard's teaching at various institutes. He subsequently spent two decades of his career in China as a geologist and paleontologist. Teilhard de Chardin spent the rest of his career outside France undertaking further scientific work. Pope Pius XII (1939–1958) opposed Teilhard. Pius's encyclical of 1950 did not rule out evolution but affirmed the ultimate authority of the church to decide in such matters. Teilhard de Chardin's view that both the physical and spiritual worlds were subject to a universal law of evolution toward a final "Omega Point" clearly went against the spirit of this encyclical. While the ban on his work has apparently never been officially withdrawn, there are many in the Catholic hierarchy who now regard his writings favorably.

Conor Cunningham's recent book, *Darwin's Pious Idea: Why the Ultra-Darwinists and Creationists Both Get It Wrong*, is a serious work of theology and philosophy.[33] He sets out in the first few chapters to debunk the view that evolution is a "theory of everything." Most scientists do not take this view, recognizing the limits of science in understanding the world. There are just a few who think that science will tell us everything and that nothing can be known apart from scientific data. Cunningham's work speaks directly to the question by arguing that most of the "extreme atheists" clearly have no understanding of theology and a poor understanding of philosophy. He also gives Dawkins's thesis of the "selfish gene" quite a pounding. Cunningham

31. Malcolm A. Jeeves and R. J. Berry, *Science, Life, and Christian Belief: A Survey of Contemporary Issues* (Grand Rapids: Baker Books, 1998).

32. H. James Birx, *Interpreting Evolution: Darwin and Teilhard de Chardin* (Buffalo, NY: Prometheus Books, 1991).

33. Conor Cunningham, *Darwin's Pious Idea: Why the Ultra-Darwinists and Creationists Both Get It Wrong* (Grand Rapids: Eerdmans, 2010).

is fascinating because he submits both "scientific creationism" and "intelligent design" to serious scrutiny and finds them wanting for similar reasons.

Cunningham accepts that evolution is God's instrument in the creative process. Evolution has now been incorporated into the theological enterprise, and many theologians see it as an integral part of describing God's character and creative activity.

Denis O. Lamoureux, professor of science and religion at St. Joseph's College at the University of Alberta, develops a Christian theology called "evolutionary creation" in his book, *Evolutionary Creation: A Christian Approach to Evolution*.[34] Lamoureux has doctorates in both biology and theology. He argues that "the Father, Son, and Holy Spirit created the universe and all life through an ordained, sustained, and design-reflecting evolutionary process."[35] Raised a Roman Catholic, he broke ranks with Christianity and became an atheist. Some years later during his military service he returned to Christianity through reading the Gospel of John. For a time, Lamoureux became a young-earth creationist. It was during his more serious theological studies that he abandoned young-earth creationism while undertaking a thorough study of the early chapters of Genesis. By the time he had completed his second doctorate in biology, he had accepted evolution as normative.

Henri Blocher was professor of systematic theology in the Faculté Libre de Théologie Evangélique at Vaux-sur-Seine, France. He wrote a study of the first chapters of Genesis, *In the Beginning: The Opening Chapters of Genesis*, which included a discussion of the creation-evolution debate.[36] Blocher, an evangelical and a conservative, concludes, "Nothing in the idea of creation excludes the use of an evolutionary procedure. Why must we tie God to one single method of action?"[37]

William P. Brown, professor of Old Testament at Columbia Theological Seminary, has closely studied a number of key biblical passages and describes seven different biblical perspectives on creation, with Genesis 1 being

34. Denis O. Lamoureux, *Evolutionary Creation: A Christian Approach to Evolution* (Cambridge: Lutterworth, 2008), xiii.

35. Lamoureux, *Evolutionary Creation*, xiii.

36. Henri Blocher, *In the Beginning: The Opening Chapters of Genesis*, trans. David G. Preston (Downers Grove, IL: InterVarsity Press, 1984).

37. Blocher, *In the Beginning*, 226.

only one of them. His close reading of the biblical text allows a synergy of understanding with scientific inquiry: "To claim the world as creation is not to denounce evolution and debunk science. To the contrary, it is to join in covenant with science in acknowledging creation's integrity, as well as its giftedness and worth."[38]

John F. Haught is professor of theology at Georgetown University and also director of the Georgetown Center for the Study of Science and Religion. In *God after Darwin: A Theology of Evolution*, he argues that Darwinian evolution provides a very fertile setting for reflecting on God: "An evolutionary theology encourages us to feel with St. Paul the Spirit of God sharing in nature's own longing for the consummation of creation."[39] It is in this spirit that Michael Dowd has attempted to integrate evolution into his evangelism in *Thank God for Evolution: How the Marriage of Science and Religion Will Transform Your Life and Our World*.[40]

Conclusion

Many scientists see no conflict between the Bible and scientific Darwinism. Oddly enough, Dawkins, a strident atheist, and creationists share a view about the incompatibility of modern science and faith.[41] The perceived conflict between science and religion, or evolution and religion, is a recent development. Since the appearance of Darwin's *Origin of Species* in 1859, most scientists and theologians have found a way to accommodate religion and science. As Peter J. Bowler points out, "such a polarized debate has few historical antecedents. In the one hundred and fifty years between 1800 and 1950, hardly any educated person would have endorsed the position we now call young-earth creationism."[42] He goes on to comment, "And within the

38. William P. Brown, *The Seven Pillars of Creation: The Bible, Science, and the Ecology of Wonder* (Oxford: Oxford University Press, 2010), 240.

39. John F. Haught, *God after Darwin: A Theology of Evolution* (Boulder, CO: Westview Press, 2000), 216.

40. Michael Dowd, *Thank God for Evolution: How the Marriage of Science and Religion Will Transform Your Life and Our World* (London: Penguin, 2009).

41. For example, see Neil Laing, *Even Dawkins Has a God: Probing and Exposing the Weaknesses in Richard Dawkins' Arguments in the God Delusion* (Bloomington, IN: WestBow Press, 2014), 6–7.

42. Peter J. Bowler, *Monkey Trials and Gorilla Sermons: Evolution and Christianity from Darwin*

evolutionists' camp there were many whose liberal views on religion allowed them to search for a way of regarding evolution as the unfolding of a divine plan established by a God whose activity is immanent within the universe."[43]

It is one thing to reveal that the scientific community, be it composed of atheists or people of faith, can assert that there is room for both evolution and God; it is another matter to understand how they reach such a conclusion. Some may suppose that acceptance of evolution might of necessity lead to atheism. How can so many Christian scientists, theologians, and biblical scholars accept the view that evolution is compatible with the Bible?

The key to answering this question hinges on a proper understanding of the Bible — how it was written and how we got it. We have already begun to see how the Bible was written as we worked through some of the early chapters of Genesis. The next chapter reviews in a more systematic way the origin of the Bible and its interpretation.

---

to Intelligent Design, New Histories of Science, Technology, and Medicine (Cambridge: Harvard University Press, 2007), 191.

43. Bowler, *Monkey Trials and Gorilla Sermons*, 191.

7

GENESIS RECONSIDERED

How We Got the Bible

Traditionally, evangelical Christians and Orthodox Jews have taken Genesis at face value; even many Muslims accept the basic story found in Genesis as true. Yet that situation is changing, perhaps more rapidly than many realize. As shown earlier, many scientists who hold an evangelical form of Christianity accept the accuracy of evolution.

The BioLogos website has a regular series on how young evangelicals are reconciling their beliefs with evolution. Two of the BioLogos staff, Kathryn Applegate and J. B. Stump, have now edited a small volume, *How I Changed my Mind about Evolution: Evangelicals Reflect on Faith and Science.*[1] In it twenty-five scientists, pastors, biblical scholars, theologians, and philosophers all give a brief autobiography of their own spiritual and scientific knowledge development. They all consider themselves evangelicals, yet they also accept evolution. One theme that cuts through many or most of these accounts is a change of understanding of the biblical text.

This reorientation among many evangelical Christians (and also Orthodox Jews) very much parallels my own personal development with regard to biblical interpretation. Originally, I had entered the academic world determined to prevent human learning and knowledge from corrupting my beliefs. I was firmly convinced of the truth of the Bible. As far as I was concerned, scholarship was valid and necessary. However, true scholarship would never conflict with the truth of the Bible.

1. Kathryn Applegate and J. B. Stump, eds., *How I Changed My Mind about Evolution: Evangelicals Reflect on Faith and Science* (Downers Grove, IL: InterVarsity Press Academic, 2016).

The first task I undertook was to investigate how we got the actual words of the Bible. In biblical scholarship this is the field of "textual criticism." It is the study of how a writer or composer gets material into written form and then how it is transmitted through the ages. Although there is a body of knowledge, the basic principles, the process, and the rules of textual criticism are used by most people in their everyday lives. Textual criticism is really "common sense" and correlates with the way most people work and think.

The first principle we need to be aware of in reading the Bible is that the writers of the Bible wrote in a particular context, with a particular background. As we saw in earlier chapters, biblical writers made use of the knowledge and patterns of thought of their times and places. This was both a strength and weakness. They drew on current knowledge, culture mores, and attitudes as they wrote for people that lived in their own day and age. For many, the actual words of the Bible are very important. They argue that the Bible is "inerrant in the autographs"; that is, the very words were true in the original writings of the biblical books. Yet these writings were written by human beings in a particular human language. We have to understand the language used by the people in the biblical world of the time to understand the meaning of the words they left for us. Some have talked of a two books view of revelation. They mean combining the Bible and insights from nature. We cannot ignore or deny the results of scientific study. If these results seem to conflict with the teachings of the Bible, we must consider whether we correctly understand the text. What I found was that my view of evolution changed not because of the science but because of the development of my understanding of the ways in which the Bible was composed and transmitted.

Languages of the Bible

Biblical writers used their own language from a particular time and place. This important point has to be reckoned with in any theology of inspiration. The Bible was not inspired in a special language. Some have assumed that divine inspiration gave the writers knowledge and understanding far beyond that of their own times and places. However, there simply is no indication in the textual remains we have that any special medium that allowed a divine message was at work. It is not the case that what we have in the manuscripts superseded the linguistic knowledge of the original writer. There is no "angelic" or "holy spirit" language. No biblical writer writes in a heavenly language, in spite of the many jokes people have coined about the language God speaks.

When knowledge of ancient Greek was rediscovered in the Renaissance, the New Testament in its original language naturally was of great interest. However, New Testament Greek was different in some respects from the language of other known Greek writings. The type of Greek in the ancient world could fairly be said to be modeled on the writings of Athenians in the fifth and fourth centuries BCE. This is what is usually called "classical Greek." The Greek language after that era underwent certain changes and developments. The Greek of the New Testament period is often referred to as "Hellenistic Greek." Some Jewish writers, such as Josephus, wrote in Hellenistic Greek. Yet the language of the New Testament remained a puzzle. Then, in the nineteenth century, a great many papyri were found in Egypt.[2] These documents were not just literary writings but also the writings of ordinary life: bureaucratic documents, legal documents, and especially the letters and other written material relating to ordinary people.

When this find of papyri was studied, scholars began to understand that the language of this "ordinary" writing resembled that of the New Testament. They realized that the New Testament was written largely in "common" Greek that was used in daily life. Young students in seminary often talk about *koine* Greek, but it is just this common Greek of the day. It was conventional for writers who were educated to use a more literary style, even to imitate classical Greek. But most of the writers of the New Testament did not have intensive education. They seem to be using the Greek language that was spoken by the common people. The New Testament was not written in a special "holy spirit" language (as was once proposed) but in the vernacular language widely spoken in the Mediterranean world.

As for the Old Testament or Hebrew Bible, much of it is composed in what is referred to as Standard Literary Hebrew. This form of the language

2. Papyrus (plural papyri) is a thin paper-like material made from the pith of the papyrus plant. Papyrus was first used in ancient Egypt but also exported throughout the Mediterranean.

was common during the later monarchy of Judah (eighth to seventh centuries BCE), as we know from inscriptions that also seem to be composed in a similar language.[3] Yet earlier forms of Hebrew do appear in the text, indicating an earlier composition. For example, the Song of Deborah in Judges 5 has been shown to be in an earlier form of Hebrew. One scholar dated the language to about the tenth or eleventh century BCE.[4] Some have also argued that Exodus 15 represents a very early Hebrew writing, exhibiting one of the earliest examples of the Hebrew language recorded.[5]

Contrasting the language of those earlier or more archaic forms of Hebrew in the Bible helps clarify that much of the Hebrew Bible is in a form of the language best known from the later monarchy. When earlier figures, such as Moses, appear in the story line, the language reveals that it was either written or revised by later scribes. Keeping in mind that the tradition, whether written or oral, was handed down in the community over many generations, it is hardly surprising that some books were composed later. The evidence that earlier traditions were used and then revised by later scribes points to their archaic origins even though written down at a later time.

Oral and Written Tradition

The present chapter is about the origins of the Bible. How did we get the Bible once it was written down? The emphasis on "written down" is deliberate because it is likely that large sections of the biblical text were first preserved by oral tradition. In agrarian and pastoral societies (such as ancient Israel), traditional or community knowledge and wisdom are passed from generation to generation by word of mouth. People tell stories about ancestors and the origins of their people. They recite community wisdom when it is needed. In some communities these traditions may also be formally taught

3. Sandra Landis Gogel, *A Grammar of Epigraphic Hebrew*, SBLSBS 23 (Atlanta: Scholars Press, 1998).

4. E. Axel Knauf, "Deborah's Language: Judges Ch. 5 in Its Hebrew and Semitic Context," in *Studia Semitica et Semitohamitica*, ed. Helen Younansardaroud, Josef Tropper, and Bogdan Burtea, AOAT 317 (Münster: Ugarit-Verlag, 2005), 167–82.

5. David A. Robertson, *Linguistic Evidence in Dating Early Hebrew Poetry*, SBLDS 3 (Atlanta: Scholars Press, 1972).

by a grandfather or grandmother to the younger generation. Some of the stories in the Bible were probably told round the hearth in the evening. Others may have had other oral contexts. It is likely that priests passed down information about the cult, the rituals, and other priestly knowledge by formal teaching.

This does not mean that writing was unavailable. Traditional agrarian societies tend to be made up of people who are largely illiterate, especially those engaged in farming, pastoral work, and labor. In such societies simply eking out a living is very labor intensive and the vast majority of people toil long to plow the fields, raise crops, or tend livestock. Most of a person's time – whether male or female, child or adult – is devoted to the task of growing and preserving food. Plowing, tending gardens or vineyards, or tending herds and flocks means work from dawn to dusk most of the year. Even so, there are times when the crops fail and food is in short supply. When all of life's energies are spent on subsistence, there is no place for books. Furthermore, the price of books would be beyond the reach of the average person, as books in the days before movable type and printing had to be copied by hand, making obtaining copies a very expensive option. Even when someone could read, there were no books readily available to read.

Recent studies have shown that most people in antiquity were not literate. One study estimated that the literacy rate in imperial Rome was 10–15 percent.[6] Catherine Hezser argues that for the Jews who lived around this same time in the first century, it was probably even lower: "the average Jewish literacy rate (of whatever degree) must be considered to have been lower than the average Roman rate."[7] The situation in ancient Israel was probably even worse. At that time, only a small group of educated and elite people were literate, including professional scribes and a few wealthy individuals.[8]

6. William V. Harris, *Ancient Literacy* (Cambridge: Harvard University Press, 1989), 328.

7. Catherine Hezser, *Jewish Literacy in Roman Palestine*, TSAJ 81 (Tübingen: Mohr Siebeck, 2001), 496.

8. See Lester L. Grabbe, *Ancient Israel: What Do We Know and How Do We Know It?* (London: T&T Clark International, 2007), 115–18. Note also Christopher A. Rollston, *Writing and Literacy in the World of Ancient Israel: Epigraphic Evidence from the Iron Age*, ABS 11 (Atlanta: Society of Biblical Literature, 2010).

Few inscriptions from the ancient Israelites have been discovered because inscriptions were not a common means of communication. Even in cultures where writing exists, a vibrant oral culture may serve to preserve the events and stories of a people. Quite sophisticated storytelling was passed on from generation to generation.[9] Such oral memory still exists in a few modern cultures. These have been extensively studied in the last century. Studying these oral cultures tells us that esoteric knowledge such as priestly lore is not written down but passed on by oral teaching. It is highly likely that large sections of the Hebrew Bible originated and were initially preserved in an oral context. However, even with its oral background, the reason the text comes to us is that these traditions were eventually written down.

Scholars debate the origin and development of the writings that came to comprise the Bible. Part of the problem is that we do not have direct information on how particular texts were composed. The best scholars can do is to infer the composition and alteration of texts from information gleaned from the books themselves or from analogies with other known literature. It is clear that books like the Psalms developed over many generations; later writers composed songs that came to be added to an existing collection. Some of the psalms allude to events many centuries after the time of David, the traditional author of the Psalms. For example, Ps 137 mentions the Babylonian captivity, which took place about 600 BCE, long after David would have lived. Other writings are composed in a late form of Hebrew that helps to date them as later compositions. For example, although the book of Ecclesiastes mentions Solomon as the literary persona of the writer, it is clear from the Hebrew language in the book, with shades of meaning only attested in later Hebrew, that it was written probably in the third century BCE, much later than the reign of Solomon.

9. Albert B. Lord, *The Singer of Tales* (Cambridge: Harvard University Press, 1960); Patricia G. Kirkpatrick, *The Old Testament and Folklore Study*, JSOTSup 62 (Sheffield: JSOT Press, 1988).

Development of the Text

Once the oral traditions migrated to the written form, what happened then? Texts in the ancient world were not static. Much like the oral traditions behind them, there was no final literary form. Texts could be added to, developed, edited, and revised. Much of the Hebrew Bible is a product of a scribal community in which many different hands have played a part in the process. In many cases, the written text reached its final form only after a long process of development. This is not universally true in the Old Testament. There are cases where we seem to have an author's own composition rather than a manuscript shaped over centuries. The book of Ecclesiastes is a case in point. Either way, copies of these ancient texts were copied by hand. A scroll might be handed down over several generations. However, they did eventually wear out, especially if read regularly.

Within these scrolls, however, was gathered material that may have had a fairly diverse origin. For instance, the books of Joshua to 2 Kings relay a history of Israel from the settlement of the land until the fall of Jerusalem to the Babylonians in 586 BCE. These books are made up of a variety of material. There are stories about ruling figures, including Joshua, the judges and Samuel, Saul, David, Solomon, and the Israelite and Judean kings, but there is much more. There are stories about various prophets. There are lists (such as David's "mighty men" and Canaanite towns taken by Israel), songs, prayers, and even a fable (Judg 9:8–15). This diversity of material was not necessarily preserved in the same way or with the same level of historical reliability.

For example, it has long been believed that the backbone of the books of 1 and 2 Kings was drawn from the Chronicles of the Kings of Israel and Judah.[10] This chronicle, mentioned in these books, was probably short with a brief entry for each king of Judah and Israel. It was likely derived from court chronicles or a court record. However, the scribes who shaped 1 Kings in some instances had an interest that went far beyond the court records. For

10. Lester L. Grabbe, "Mighty Oaks from (Genetically Manipulated?) Acorns Grow: The Chronicle of the Kings of Judah as a Source of the Deuteronomistic History," in *Reflection and Refraction: Studies in Biblical Historiography in Honour of A. Graeme Auld*, ed. Robert Rezetko, Timothy Lim, and W. Brian Aucker, VTSup 113 (Leiden: Brill, 2006), 154–73. I argue that there was only one chronicle used (Chronicle of the Kings of Judah) that included material not only on Judah but also on Israel.

example, the "history of Ahab" (1 Kgs 16:29–22:40) includes stories about the prophets Elijah, Elisha, Micaiah, and others. There is much more color in the story of Naboth's vineyard and the narrative of the battle in which King Jehoshaphat and King Ahab fight a common enemy, a battle that results in the death of Ahab. All of these details are from other sources than court records and tell us something of the interest and context of the writers or compilers of the book of 1 Kings.

The situation is similar with the Pentateuch (the five books attributed to Moses: Genesis, Exodus, Leviticus, Numbers, and Deuteronomy); the traditions are mixed. There are passages in the Pentateuch itself that suggest that Moses may have written the books or parts of them. However, the linguistic analysis of the books demonstrates that the Hebrew bears the marks of a time much later than that of Moses. Furthermore, the book of Genesis is set long before Moses, even though it is written in Standard Literary Hebrew, not an early dialect. Moses's own death is described in Deuteronomy 34:5–8, something he could not have accomplished as the sole author. Some have appealed to Mark 7:10 to "prove" that Moses wrote the books. However, the practice at the time was to refer to standard attributions without dispute. We find the same phenomena in the book of Jude, which quotes from the Book of Enoch as if it was the patriarch Enoch (see Jude 14–15). However, we know that the words in Enoch were not actually uttered by Enoch but by Jewish scribes of the third century BCE.

There is, in fact, evidence that the Pentateuch was not compiled until the Persian period, probably in the decades after 400 BCE, though some of the material in it is much older. The books of Ezra and Nehemiah refer to the book that Ezra brought to Palestine and read to the people. Many scholars believe this refers to the Pentateuch (Ezra 7; Neh 8–9). It is clear from these passages that the book read by Ezra and others was new to the people. The Jewish community on the island of Elephantine in Egypt is a literary trove where many letters and other documents were found, dating from the fifth century BCE. Some of these documents are actually dated around 410 BCE. Some documents mention Jerusalem, the high priest Yohanan,[11] and the nobles of the people. The Jewish community on

11. Yohanan (about 400 BCE) was the fifth high priest after the rebuilding of the temple in Jerusalem by the Jews after the Babylonian captivity (cf. Neh 12:22–23).

Elephantine was in contact with Jerusalem and was aware of the temple and priesthood there. They also kept certain standard Jewish observances, such as Passover and the Sabbath.[12] Yet in all their literature, religious and legal, there is not one reference to Moses, Aaron, or the Pentateuch. They did not know the Pentateuch as we have it, even though they knew about some of the traditional Jewish practices described in it. Some date the coming of Ezra to Jerusalem with the Pentateuch about 400 BCE or a bit later.

However, around 200 BCE a Jewish scribe named Ben Sira compiled a list of Jewish writings regarded as having religious or spiritual authority. The book – Sirach, as it is known in Greek, or Ecclesiasticus in the Latin sources – lists most (not all) of the books of our present Hebrew Bible (Sir 44–49). It is clear that the Pentateuch is an integral part of Ben Sira's Bible. This demonstrates that between the time of the Elephantine community and this time (around 200 BCE), the books of the Pentateuch had been compiled, edited, and established as religious writing having authority for the Jews. This is not yet what would become the standard canon of the Hebrew Bible.

The Transmission of the Text

Since the invention of the printing press, readers have become accustomed to the idea that the original text is reproduced faithfully, as many times as needed, exactly as the original. The normal expectation is that the original text has been carefully proofread and is as free from error as possible. Careful readers also know that every printed book or paper has errors and imperfections. If the printed or now digital text can reflect human error, how much more so was the case when each copy had to be done by hand! Each copy contains some mistakes. Words might be omitted, duplicated, or changed, either intentionally or unintentionally. These are normal human failings that regularly occur when copying a text by hand. Of course, having copied

12. See Lester L. Grabbe, "Elephantine and the Torah," in *In the Shadow of Bezalel: Aramaic, Biblical, and Ancient Near Eastern Studies in Honor of Bezalel Porten*, ed. Alejandro F. Botta, CHANE 60 (Leiden: Brill, 2013), 125–35.

a section of text, we can then check it and correct any mistakes by inserting the missing words between the lines or in the margin. This was normally the way copyists fixed their mistakes in antiquity. Many surviving examples of manuscripts exist with such marginal corrections made by either the original copyist or by someone else who corrected the manuscript at a later time.

The finds at Qumran, better known as the Dead Sea Scrolls, provide ample evidence of the copying of textual materials and the sorts of errors they contain. There are many biblical manuscripts among the scrolls found near the Dead Sea. They give us an understanding of how the written text was passed down by scribes copying by hand over hundreds of years. The scrolls reveal how mistakes were corrected, how damaged manuscripts were repaired, and how text damaged by use was made legible once more. If others wanted to use them, they either had to have access to the manuscript(s) of the community or they had to have a copy made for their own use. Copies were proofread for mistakes and inadvertent changes to the text, but a certain number always slipped through. As texts were copied and moved to new locations, the result was that over time slightly different texts were being used in different geographical areas. Even the best manuscript copies differed at least slightly from each other, and sometimes the differences were much greater.

Diverse Textual Traditions

The various books of the Bible went through a process of writing and alteration over a period of time before reaching their final form. Many biblical books were continuing to grow and develop in the Greek and Roman periods. The biblical manuscripts found at Qumran and elsewhere often have two or more versions of the biblical text. The book of Jeremiah is noteworthy because there were at least two significantly different versions in circulation. There is the traditional Hebrew text of Jeremiah known as the Masoretic text. This is also the so-called long version of Jeremiah. However, the Septuagint version of Jeremiah is one-sixth shorter than the Hebrew Masoretic (traditional Jewish text) and has certain sections in a different order. The difference in length is not the simple result of translation. At Qumran we find a Hebrew version of Jeremiah that coincides with the Greek.

The Greek of Jeremiah was therefore translated from a Hebrew manuscript that was different from the Hebrew Masoretic. Many scholars argue that the shorter version of Jeremiah is earlier and more original.

Other parts of the Bible show a similar diversity in textual evidence. The book of Joshua in Greek is quite different in places from the Hebrew text. Which is more original? This is hard to say, since differences in the Greek text could be the result of an originally different Hebrew text. The same applies to Job, which is one-eighth shorter in Greek. Is this a result of translation or is it because of a different Hebrew text than the Masoretic? The problem is that not many early Hebrew manuscripts of Job have survived. The Septuagint Greek text has a long additional section in 1 Kings 12:24 that is not in the Masoretic Hebrew. The books of Esther and Daniel have multiple versions when the various Greek versions are taken into account. Some of this diversity may be a result of translation, but some of it seems to stem from a different "original" Hebrew text.

The Pentateuch and the books of Samuel demonstrate a remarkable diversity of versions. There is manuscript evidence for at least three different versions of each. For the Pentateuch, we have the two Hebrew versions: (1) the Masoretic text in Hebrew used by the Jewish community (MT) and (2) the Hebrew text used by the Samaritan community (SP). In addition, we have the Septuagint text in Greek (LXX), which is likely translated from a Hebrew text differing from both the Masoretic text and the Samaritan Pentateuch. Furthermore, for the book of Genesis we have a paraphrase in early Jewish writing known as Jubilees (probably from the second century BCE). The following table illustrates the point. This is the genealogy found in Genesis 5 and subsequent chapters (esp. Gen 11), with the different ages of the patriarch in question when he sired a son.

Table 2: Comparison of Genesis 5 in Different Textual Traditions

Biblical Text	SP	LXX	MT	Jubilees
5:3: Adam begets Seth.	130	230	130	130
5:6: Seth begets Enosh.	105	205	105	98
5:9: Enosh begets Kenan.	90	190	90	97

Biblical Text	SP	LXX	MT	Jubilees
5:12: Kenan begets Mahalalel.	70	170	70	70
5:15: Mahalalel begets Jered.	65	165	65	66
5:18: Jared begets Enoch.	62	162	162	61
5:21: Enoch begets Methuselah.	65	165	65	65
5:25: Methuselah begets Lamech.	67	167	187	65
5:28: Lamech begets Noah.	53	188	182	49–55
5:32; 11:10: Noah begets Shem.	502	502	502	500–506
7:11: Noah's age when flood came.	600	600	600	601–607
11:10: Shem's age at the time of the flood.	98	98	98	101
11:10: Shem begets Arpachshad.	100	100	100	103
11:12: Arpachshad begets Shelah (begets Cainan in LXX and Jubilees).	135	[135]	35	[65]
11:13: Cainan begets Shelah.		[130]		[57]
11:14: Shelah begets Eber.	134	130	30	71
11:16: Eber begets Peleg.	134	134	34	64
11:18: Peleg begets Reu.	130	130	30	12 [sic]
11:20: Reu begets Serug.	132	132	32	108
11:22: Serug begets Nahor.	130	130	30	57
11:24: Nahor begets Terah.	70	79 [or 179]	29	62
11:26: Terah begets Abram (11:32; 12:4).	70	130	130	70
17:17; 21:5: Abraham begets Isaac.	100	100	100	112
25:26: Isaac begets Jacob.	60	60	60	58

As is plain here, these different ancient texts have four different genealogies with timelines that do not correspond. Some will argue that we should simply follow the Hebrew. While this would be tidy, it is overly simplistic. The New Testament often quotes the Septuagint text as the authoritative version even when it differs from the Hebrew. The genealogy of Jesus quoted in the gospel of Luke (3:36) has an extra person in it, the person Cainan. This figure is missing from Genesis 5:12–13 in the Hebrew text but is found in the Septuagint and in the book of Jubilees, the original text of which was in Hebrew. These different versions were relevant and authoritative in the first century CE, the time of Jesus and the birth of Christianity.

For those who argue that the Bible is the basis for determining the age of the earth, these differences pose significant problems. How would one know that any of these four represents the original version? The complex history of the text raises the possibility that there may have been another version — no longer preserved — with quite a different genealogy and quite a different chronology.

Similarly, the books of Samuel survive in two Hebrew versions, one in the Masoretic text and one in the Hebrew text from Qumran (though only preserved in fragments).[13] There is also a Greek translation. The differences are not great, but they do vary in detail. For example, 1 Samuel 1:11 in the traditional Masoretic Hebrew text (the text translated in most versions) says about the child promised to Hannah, "And I shall give him to YHWH all the days of his life, and a razor shall not go upon his head." The Greek version has more detail: "And I shall give him to you [God] as a gift until the day of his death, and wine and strong drink he shall not drink, and a razor shall not go upon his head." The Hebrew text from Qumran seems to read: "And I shall give him to you as a Nazarite until the day of his death, and wine and strong drink he shall not drink, and a razor shall not go upon his head." There are other such variants among the versions of the books of Samuel as well.

The nature of the textual evidence from which we reconstruct the text for modern translations problematizes the doctrine of inerrancy, the idea that the Bible is without error in the original autographs. Yet how can we speak of inerrancy of the original text when there is no original text to consult? The doctrine of inerrancy would require that God had acted to preserve a single authoritative text. That is simply not the way we got the Bible today. The actual textual evidence available does not always allow us to say with certainty what the original text was. For the Bible to be "literally true"

13. For the Qumran fragments, see Frank Moore Cross, Donald W. Parry, Richard J. Saley, and Eugene Ulrich, *Qumran Cave 4.XII: 1–2 Samuel*, DJD 17 (Oxford: Clarendon, 2005).

requires that we have a text that is uncompromised and whose meaning is not ambiguous. Neither is always the case. In fact, the writers of the books that came to comprise the New Testament were quite happy to quote a variety of versions of biblical texts with no qualms about authority.

Quotations of the Hebrew Bible in the New Testament

Many of those who hold a view of inerrancy would be astonished to realize how unconcerned the writers of the New Testament often were about quotations from and references to the Old Testament. They seldom quoted the Hebrew text or even a literal translation of the Hebrew text. These authors frequently cited Greek translations. Yet, we also find other versions used quite freely. There are even instances where the writers got the attributions wrong. For example, in Matthew 27:9 the quotation is assigned to "the prophet Jeremiah." However, the quotation is actually from Zechariah 11:13. We also have Jewish writings quoted as authoritative that are not in the Jewish canon. The outstanding example is the quotation from the book of 1 Enoch in Jude. For many centuries a Greek translation of this passage was available. It was close to the quotation in Jude but not exact. However, when an Aramaic copy of 1 Enoch 1:9 was found among the Dead Sea Scrolls, it turned out that Jude was even closer to the original than the Greek translation of 1 Enoch:

Table 3: Comparison of 4QEn[c], Greek Fragments, and Jude 14–15

4QEn[c]	Greek Fragments	Jude 14–15
[כדי יאתה]	ὅτι ἔρχεται	Ἰδοὺ ἦλθεν κύριος
[עם רבו] את קדיש [הי]	σὺν ταὶς μυριάσιν ἀυτοῦ καὶ τοῖς ἁγίοις ἀυτοῦ,	ἐν ἁγίαις μυριάσιν αὐτοῦ,
[למעבד דין על כולהון ויובד כול רשיעין]	ποιῆσαι κρίσιν κατὰ πάντων, καὶ ἀπολέσει πάντας τοὺς ἀσεβεῖς,	ποιῆσαι κρίσιν κατὰ πάντων,

4QEn[c]	Greek Fragments	Jude 14–15
[ויוכה ב]שרא על עובד[י] רשעהון כולהון	καὶ ἐλέγξει πᾶσαν σάρκα περὶ πάντων ἔργων τῆς ἀσεβείας αὐτῶν	καὶ ἐλέγξαι πᾶσαν ψυχὴν περὶ πάντων τῶν ἔργων ἀσεβείας αὐτῶν
[די עבדו ומללו לארשעה]	ὧν ἠσέβησαν	ὧν ἠσέβησαν
ועל כול מלין] רברבן וקשין [די מללו עלוהי חטין רשיעין]	καὶ σκληρῶν ὧν ἐλάλησαν λόγων, καὶ περὶ πάντων ὧν κατελάλησαν κατ' αὐτοῦ ἁμαρτωλοὶ ἀσεβεῖς.	καὶ περὶ πάντων τῶν σκληρῶν ὧν ἐλάλησαν κατ' αὐτοῦ ἁμαρτωλοὶ ἀσεβεῖς.
[When he comes	Because he will come	Behold, the Lord has come
with] the myriads of his holy ones,	with his myriads and his holy ones	with myriads of his holy ones
[to execute judgment against all, then he will destroy all the wicked]	to execute judgment against all and destroy all the wicked	to execute judgment against all
and convict] flesh of [all their] works [of wickedness	and he will convict all flesh of all their wicked deeds	and to convict every soul of all their wicked deeds
which they did and said,	which they have done wickedly and harshly.	that they have committed in such an ungodly way,
and of all] the proud and hard [words which wicked sinners have spoken against him].	And (the hard) words which they will speak, and concerning all that ungodly sinners have spoken against him.	and of all the harsh things that ungodly sinners have spoken against him.

This simple table demonstrates that the writer of Jude was not overly concerned with the exact words of the cited text. The important thing was the overall message. In addition to free adaptation of sources, the writers of the New Testament employed means of interpretation that were common in their day, even though these would be rejected by modern scholars.[14] The authors of the Bible were indebted to their environment and contemporary culture.

Conclusion

The Bible consists of human literature. It was written, copied, translated, and transmitted by human beings. This is not a denial of divine inspiration, but it means that the Bible did not fall from heaven in 1611 in a hermetically sealed package, printed by the finger of God, complete with red lettering, gold edging, leather covers, and silk marker ribbons. In whatever way God was behind the producers of the Bible – however the originators were inspired – it was still made by human beings, in their language and according to conventions of their literature, with all their limitations of knowledge and expression. Contemporary study suggests that large portions of the Bible originated in an oral context and were passed down orally for a shorter or longer period of time. Oral literature had a large place in traditional communities before the modern period and was often rich and sophisticated. The conventions of creating and passing it on are now much better understood and appreciated.

The literature that later went on to become a part of the Bible was primarily community literature. Our modern experience is that a publication has a single author or perhaps a couple of authors. Even anonymous writings are often the product of a single or perhaps a couple of authors. Multiple authorship is fairly common in some scholarly writings, especially scientific papers. Also, many papers that reach print have undergone an editorial process that has involved one or more editors (in addition to the authors). Even here, though, we do not encounter a literary product from a community. Ancient literature was often a text with many contributors that had been handed down over a number of generations. Some single-authored works have probably made it into the Bible, such as the book of Qohelet or Ecclesiastes, but these are the exception.

14. For further on this subject, see Matthias Henze, ed., *A Companion to Biblical Interpretation in Early Judaism* (Grand Rapids: Eerdmans, 2012).

Finally, the literature reflects the times of the composers and writers. When Hebrew was their language, they wrote in Hebrew. When Aramaic became a part of the linguistic scene, we find Aramaic (as in Ezra and Daniel). Later, we find Jewish literature written in Greek. The Hebrew Bible was translated into Greek. Indeed, it was the Greek version that became the Bible of the early church.

It is very important to recognize that the composers of the text were very much creatures of their own time and world. They used their contemporary language, not some ethereal celestial language or "tongues of angels." They were limited by their own environment and knowledge. They spoke about what they knew. Just as their language was the language known and used by those around them, so their knowledge of the world reflected the knowledge of the times. At times we see evidence of a more sophisticated, learned knowledge of the natural world, but it was knowledge attested in other literature of the time. There is no evidence that the ancient writers had insight into or knowledge of modern physics, biology, or astronomy. Just as we would not expect them to know modern English, we should not expect them to understand modern genetics or geology.

Plate 1. The Ancient Hebrew Conception of the Universe © Michael Paukner

Plate 2. *Pakicetus* © Roman Uchytel

Plate 3. *Ambulocetus natans* © Roman Uchytel

Plate 4. *Kutchicetus* © Roman Uchytel

Plate 5. *Maiacetus* © Roman Uchytel

Plate 6. *Basilosaurus* © Roman Uchytel

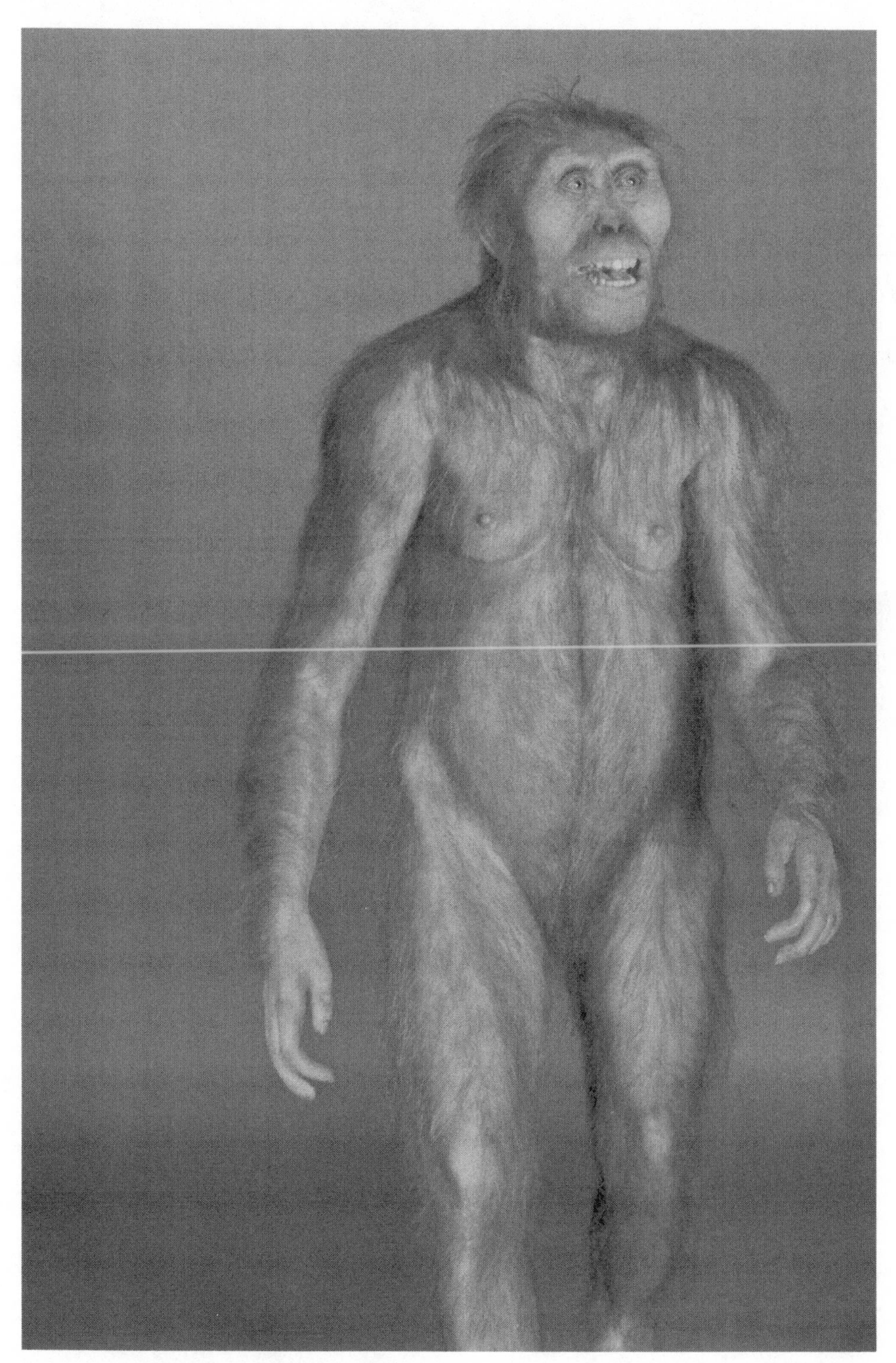

Plate 7. Lucy © Elisabeth Daynes / LookatSciences
Reconstruction: Elisabeth Daynes, Paris

Plate 8. *Homo erectus* © S. Entressangle, E. Daynes / LookatSciences
Reconstruction: Elisabeth Daynes, Paris

Plate 9. *Homo neanderthalensis* © Elisabeth Daynes / LookatSciences
Reconstruction: Elisabeth Daynes, Paris

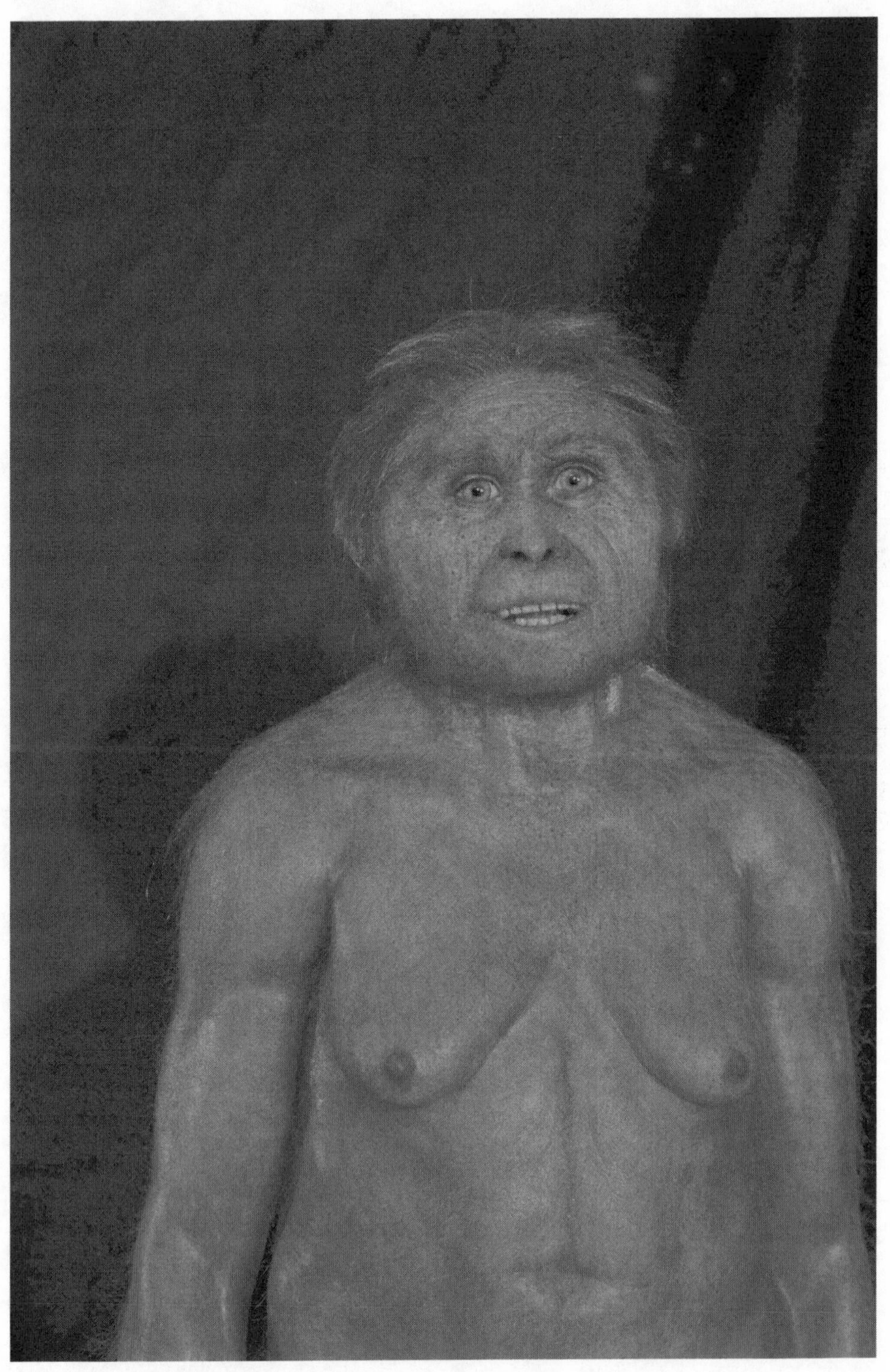

Plate 10. *Homo floresiensis* © Elisabeth Daynes / LookatSciences
Reconstruction: Elisabeth Daynes, Paris

PART III

Adam and Human Ancestry

8

HOMO NEANDERTHALENSIS, HOMO DENISOVA, AND *HOMO SAPIENS*

Deep Ancestry

The understanding of human history, our deep ancestry, is changing rapidly. Scientists past and present have been hard at work on the challenging task of interpreting the data and constructing new explanatory theories and exploratory hypotheses.

The Origin of Primates

Mammals first arose during the Triassic period, early in the Mesozoic era.[1] The dinosaurs became dominant during the next two periods, the Jurassic and the Cretaceous. At that time, mammals were small. However, when the dinosaurs died out at the end of the Cretaceous period, mammals began to fill the various ecological niches in the first epoch (Paleocene) of the new Cenozoic era. The earliest primate fossils are found in the next epoch, the Eocene (c. fifty-five to thirty-four million years ago). DNA studies suggest that primates might have originated as early as the Cretaceous period when the dinosaurs were still around. Primates make up the mammalian class that includes lemurs and lorises, monkeys, marmosets, baboons, gibbons, orangutans, gorillas, chimpanzees, and humans.

1. See Robert Boyd and Joan B. Silk, *How Humans Evolved*, 7th ed. (New York: Norton, 2015); John G. Fleagle, *Primate Adaptation and Evolution*, 3rd ed. (San Diego: Academic Press, 2013); Chris Stringer and Peter Andrews, *The Complete World of Human Evolution* (London: Thames and Hudson, 2005); Richard G. Klein, *The Human Career: Human Biological and Cultural Origins*, 3rd ed. (Chicago: University of Chicago Press, 2009).

The earliest primate fossil dates to approximately fifty-five million years ago. It is a hand-sized creature (which apparently looked somewhat like a modern tarsier), the *Archicebus achilles*.[2] The fossil "Ida," *Darwinius masillae* (c. forty-seven million years ago), is preserved in detail with fur intact. Although some debate exists about its classification, a number of paleoanthropologists place it among the primates.

The Miocene epoch extended from about twenty-three to five million years ago. Gorillas, chimpanzees, and humans are widely attested in this period. Genetic studies have suggested that the last shared ancestor for humans, gorillas, and chimpanzees lived eight or nine million years ago, while for humans and chimps it was five to eight million years ago.[3] Hominins, our humanlike ancestors, developed in the Miocene epoch. *Sahelanthropus tchadensis*, quite possibly the earliest hominin fossil (six to seven million years ago), was found near Lake Chad in Africa. This creature is notable because it was apparently bipedal; that is, it walked upright on two legs. This conclusion is based on the fact that the head was apparently balanced on top of the spine. Unfortunately, only the cranium has been found. The teeth and cranial structure have some features that are more like humans than apes, yet its brain was only the size of a chimpanzee's.

Most hominins in this period belong to the genus of *Australopithecus* and are often referred to collectively as "australopithecines" or "australopiths." One of the most famous is popularly known as "Lucy," a female australopithecine (*Australopithecus afarensis*) dated to more than three million years ago, much of whose skeleton was preserved. Most paleoanthropologists conclude that she walked on two legs when on the ground. On the other hand, her long arms, short legs, and curved fingers suggest that she spent a good amount of time in the trees. In fact, recent study of the remains suggests that she may have died from falling from a tree. Although she evidently walked upright, her anatomy is somewhat different from modern humans. She may have walked differently. Her brain size was about the same as that of chimpanzees. The hyoid bone of an australopithecine child

2. Xijun Ni et al., "The Oldest Known Primate Skeleton and Early Haplorhine Evolution," *Nature* 498 (2013): 60–64.

3. Boyd and Silk, *How Humans Evolved*, 233.

has been preserved and indicates that the species did not possess the ability for speech.[4] (See Lucy on Plate 7.)

The "Laetoli footprints," a set of footprints thought to have been made by several *Australopithecus afarensis* individuals walking through wet volcanic ash about three and a half million years ago, indicate that those who made them were walking bipedally.

Australopithecines resembled modern humans by walking upright and having some similar teeth. However, in other ways, they tend to exhibit the physical characteristics and life cycle of great apes.[5] More proximate ancestors of modern humans seem to have developed in the period of about six to two million years ago: "Although there are many gaps in the fossil record and much controversy about the relationship among the early hominin species, we do know several important things about them. In every plausible phylogeny [family tree], the human lineage is derived from a small biped who was adept in trees. Its teeth and jaws were suited for a generalized diet. The males were considerably taller and heavier than the females, their brains were the same size as those of modern apes, and their offspring developed faster than modern humans do."[6]

Members of the Genus Homo

The discovery of fossil remains in the Olduvai Gorge and elsewhere in eastern Africa led to the identification of *Homo habilis*, a being who lived two and a half million years ago. Among the fossils, some were more robust than others. Some paleoanthropologists want to see two species, the less robust being *Homo habilis*, with more humanlike teeth and a smaller brain (about 500 cc), and the more robust *Homo rudolfensis*, with massive teeth and jaws and a larger brain (about 770 cc). Whether two species or only one, the size of these brains is between that of chimpanzees and modern humans. They seem to have mastered the art of making stone tools. A recent article in *Nature* argued that stone blades were being made around three and a

4. Boyd and Silk, *How Humans Evolved*, 251–52.

5. Klein, *Human Career*, 132, 195–96.

6. Boyd and Silk, *How Humans Evolved*, 264.

half million years ago, before any of the *Homo* genus existed. This suggests what others had already concluded: australopithecines were making tools.[7] In any case, the use of tools by *Homo habilis* earned the species its name, "handy man."

Another important human technology was the use of fire. Fire afforded many benefits, including a means to live in an inhospitably cold climate. Exactly when humans developed control of fire is much debated. *Homo erectus* and *Homo ergaster* may have used fire about a million years ago. Although *Homo neanderthalensis* used fire, they did not seem to exploit it to the extent that some other hominins did. *Homo neanderthalensis* evolved a short, stocky body that allowed them to resist the cold. *Homo sapiens* had a different body shape. They used fire and clothing to keep warm. A recent study has found that *Homo sapiens* developed a gene that permitted the toleration of smoke and the products of fire in a way that differed from other species of humans.[8] (See *Homo erectus* on Plate 8.)

The next human development was *Homo erectus*, roughly two million years ago. Australopithecines continued to exist for another million years, but this is not surprising: species do not disappear just because a new one develops. Scientists believe that *Homo erectus* developed from *Homo habilis*. Some paleoanthropologists see two slightly different versions of *Homo erectus*. However, the trend is to refer to the African version as *Homo ergaster*, while those from Asia are called *Homo erectus*. *Homo erectus/ergaster* is recognized by most specialists as the first human.

Some representatives of *Homo ergaster* migrated into Eurasia about one million years ago, leading to the humans referred to as *Homo erectus* in China and other parts of Southeast Asia. These humans had large brains (c. 875 cc) and walked upright. Although very similar to modern humans, *Homo erectus/ergaster* differed from modern humans. Their skull shape was long and low. They had brow ridges. They may have used fire, though this is still uncertain.[9] They do not seem to have been hunters or scavengers, though information on this is difficult to ascertain from human and animal

7. Sonia Harmand et al., "3.3-Million-Year-Old Stone Tools from Lomekwi 3, West Turkana, Kenya," *Nature* 521 (2015): 310–15. See also Klein, *Human Career*, 249.

8. Troy D. Hubbard et al., "Divergent Ah Receptor Ligand Selectivity during Hominin Evolution," *Molecular Biology and Evolution* 33.8 (2016).

9. Klein, *The Human Career*, 412–14.

remains.[10] There were regional differences between those of the same human species. The new environment in many cases encouraged conservatism or change in various physical features. There are arguments that they began to develop speech, based on anatomy and other considerations, though this remains speculative. By half a million years ago our human ancestors had spread widely across Africa, Asia, and Europe.

A human once referred to as ancient *Homo sapiens* appeared more than half a million years ago and is now classified as *Homo heidelbergensis*. It had much in common with modern humans, including a large brain (average c. 1250 cc), not much smaller than the average person today. This species is thought to be the ancestor not only of *Homo sapiens* but also *Homo neanderthalensis* and *Homo denisova*. Scientists think *Homo heidelbergensis* evolved in Africa from *Homo ergaster*. *Homo heidelbergensis* migrated into Europe and Asia because fossils of it have been found in Israel, Germany (the first *Homo heidelbergensis* fossil in 1907), France, England, Spain, and Greece. *Homo neanderthalensis* seems to have evolved from individuals who migrated into Europe, and *Homo denisova* from those who moved into the central part of Asia. *Homo heidelbergensis* differed from modern humans primarily in the size of its brain and the presence of brow ridges. (See *Homo neanderthalensis* on Plate 9.)

Homo neanderthalensis, first discovered in 1856, has been extensively studied. At one time, *Homo neanderthalensis* was classified as *Homo sapiens neanderthalis*. However, most recently they have been identified as a separate species (*Homo neanderthalis*). Their average brain size (c. 1530 cc) was even larger than that of modern humans (c. 1400 cc), though there is no indication that they were more intelligent. Their large brains may have been the result of a body structure that was much heavier, stockier, and better muscled (though slightly shorter) than the average *Homo sapiens*. The body seems to have evolved to survive in the cold, since they lived in Europe during glacial conditions. *Homo neanderthalensis* does not seem to have built shelters, though they made use of caves and rock overhangs. They appear to have buried their dead.

In 2010 at the site of the Denisova Cave, Siberia, fragments of finger bones were found, followed by some teeth. They date from thirty to fifty thousand years ago. DNA was extracted from the fossils, which enabled the sequenc-

10. Klein, *The Human Career*, 416–23.

ing of a genome. The fossils were judged to be of an early human related to but different from *Homo neanderthalensis*. This is the first human to be reconstructed on the basis of DNA rather than fossils, but scientists are confident that this was another early human group. However, because the exact identification could not be made on the basis of fossil evidence, the people have been called "Denisovans."[11] *Homo denisova* and *Homo neanderthalensis* apparently diverged into separate branches about four hundred thousand years ago.

One of the most curious humanlike creatures is *Homo floresiensis*, who were announced to the world in 2004.[12] Nicknamed "hobbits" because of their small size (only about a meter tall), they lived on the island of Flores in Indonesia. Although initial finds suggest they lived eighteen to sixteen thousand years ago, they were different from modern humans. Their brain size was only about 400 cc. They had arms proportionally longer than modern people. *Homo floresiensis* also had large feet with no longitudinal arch. Some scientists have argued they were modern humans suffering from some sort of malady, for example, Down syndrome.[13] The dating has also been reassessed, with the argument now that the remains should be dated sixty to one hundred thousand years ago; the stone tools associated with the skeletal remains are now dated from fifty to one hundred and ninety thousand years ago.[14] (See *Homo floresiensis* on Plate 10.)

In 2016, older material was found at a different site on Flores, including stone tools, part of a jaw, and teeth from three hominins that date to about seven hundred thousand years ago.[15] Although there is some uncertainty, because no skulls were found, the jaw indicated an individual who was even smaller than the "hobbits" first found. Because one of the teeth is a wisdom tooth, the jaw is most likely from an adult. If the new finds are definitely from *Homo floresiensis*, they seem to disprove the idea that "hobbits" were

11. See David Reich et al., "Genetic History of an Archaic Hominin Group from Denisova Cave in Siberia," *Nature* 468 (2010): 1053–60.

12. For information on this species, see Chris Stringer, *The Origin of Our Species* (London: Penguin Books, 2011), 78–82; also Boyd and Silk, *How Humans Evolved*, 313–14.

13. Karen L. Baab et al., "A Critical Evaluation of the Down Syndrome Diagnosis for LB1, Type Specimen of *Homo floresiensis*," *PLOS One* (2016): 1–32.

14. Thomas Sutikna et al., "Revised Stratigraphy and Chronology for *Homo floresiensis* at Liang Bua in Indonesia," *Nature* 532 (2016): 366–69.

15. Gerrit D. van den Bergh et al., "*Homo floresiensis*-like Fossils from the Early Middle Pleistocene of Flores," *Nature* 534 (2016): 245–48.

modern humans with a growth disorder. Another puzzle is how the tools associated with them seem so advanced. Their brains do not appear to be large enough for them to have developed such technology. But a study on dwarfism among island populations has indicated that brains of such individuals shrink even further than their bodies.[16]

An even more recent find is the creature that has been labeled *Homo naledi*. Human remains were found in a cave that is difficult to access in the Cradle of Humankind World Heritage Site in South Africa in 2013. They were excavated, and the preliminary results of the study were released in September 2015.[17] The physical characteristics of the individuals were somewhat puzzling. A number suggested that they should be classified as *Homo*: the shape of the skull, jaw, teeth, and spine; bipedal locomotion, though the foot was not fully developed toward human shape; and the wrists and hands were like those of humans and shaped for object manipulation. However, there were also features more like an australopithecine: a small brain (only c. 500 cc); shape of the pelvis, shoulders, and rib cage; and long curved fingers suitable for moving in trees. Although the fifteen or so individuals found had much in common with the genus *Homo*, the small brain (only about half the size of *Homo erectus*) was the main obstacle. Thus, some are not convinced it should be classified as *Homo*, while others think it should probably be labeled *Homo erectus*. No doubt further study will lead to reevaluation of many of the preliminary results.

Finally we come to our own species, *Homo sapiens*, apparent descendants of *Homo heidelbergensis*. *Homo sapiens* evolved in Africa about two hundred thousand years ago. According to the most recent theory the distribution of this group came about when they migrated from Africa about fifty thousand years ago and, over time, populated Europe, Asia, and then from Asia down to Australia and across to North and South America. At this time (c. fifty thousand years ago) a major change seems to have taken place. The result was a considerable advance in culture and technology. Although the brain

16. Eleanor M. Weston and Adrian M. Lister, "Insular Dwarfism in Hippos and a Model for Brain Size Reduction in *Homo floresiensis*," *Nature* 459 (2009): 85–88.

17. Lee R. Berger et al., "*Homo naledi*, A New Species of the Genus Homo from the Dinaledi Chamber, South Africa," *eLife* (2015); Paul H. G. M. Dirks et al., "Geological and Taphonomic Context for the New Hominin Species *Homo naledi* from the Dinaledi Chamber, South Africa," *eLife* (2015).

stayed the same size, some have proposed that a significant evolutionary development — a major restructuring of the brain — took place. This cannot be demonstrated from the fossil crania found but is inferred from the cultural advance. Others, however, are skeptical of such a physical change, though they accept the drastic upgrading of the technology.

There is a growing consensus that modern humans emerged out of Africa as opposed to independent development in different parts of the world from earlier hominins that had settled in that particular region. One of the early representatives of *Homo sapiens* was Cro-Magnon in Europe about thirty thousand years ago. A number of scientists think that Cro-Magnon people drove *Homo neanderthalensis* to extinction. That is not necessarily the case. A change in climate and resources made *Homo neanderthalensis* more vulnerable. Others think that interbreeding with *Homo sapiens* was the cause of *Homo neanderthalensis* demise. In any case, they evidently died out just as *Homo sapiens* made its cultural leap.

Modern Humans Developed from Closely Related Primates

Studies show that all primates are related. Great apes are very closely related to humans.[18] This is demonstrated not only by shared physiology — known from anatomical studies at least since the nineteenth century — but especially through more recent genetic analysis. It is not just that 96 to 99 percent of DNA is shared between humans and chimpanzees but also that the pattern of shared genetics makes it almost impossible to be the result of accident or coincidence. The pattern of shared genes between creatures with common ancestry is now widely recognized; the development of primates from an ancestor common to humans and the other great apes can be traced without major difficulty to between eight and five million years ago.

Evidence of evolution exists in "pseudogenes," genes that were formerly active but have become inactive because of mutation. They continue to be duplicated and passed down as part of the chromosome DNA. A good exam-

18. For example, Asper Hobolth et al., "Genomic Relationships and Speciation Times of Human, Chimpanzee, and Gorilla Inferred from a Coalescent Hidden Markov Model," *PLoS Genetics* 3.2 (2007): 294–304.

ple is the pseudogene for L-gulonolactone oxidase, which apparently once allowed all primates to produce their own vitamin C.[19] At some point early in primate ancestry, the gene became inactive in the Haplorhini branch of the primate line (which included monkeys, apes, and humans), though sufficient vitamin C was being ingested in the diet that this caused no problem. Therefore, the loss of this gene did not cause the line to die out by natural selection. This is why humans today must obtain sufficient vitamin C in their diets, even though some other animals retain the ability to manufacture it in their bodies. But this once-active gene has become a pseudogene in the entire primate line. This is only one pseudogene, but there are many more and their pattern in the genome allows us to see that the various primates are related and also how closely.

Evangelical professor of biology at Trinity Western University in Langley, British Columbia, Dennis Venema has written extensively on evolution and genetics.[20] As he demonstrates, genetics show that modern people did not descend from a single primal pair of humans like Adam and Eve about six thousand years ago. He points out, "Taken individually and collectively, population genomics studies strongly suggest that our lineage has not experienced an extreme population bottleneck in the last nine million years or more . . . and that any bottlenecks our lineage did experience were a reduction only to a population of several thousand breeding individuals. As such, the hypothesis that humans are genetically derived from a single ancestral pair in the recent past has no support from a genomics perspective, and, indeed, is counter to a large body of evidence."[21]

On the contrary, we come from a population of probably about six to ten thousand individuals who lived roughly one hundred thousand years ago. A "bottleneck" is a situation in which the population became very small. If there were a single couple who gave life to all humans alive today, this would be an extreme bottleneck. Yet careful consideration of the genomics demonstrates that this simply did not happen.

19. See Guy Drouin, Jean-Rémi Godin, and Benoît Pagé, "The Genetics of Vitamin C Loss in Vertebrates," *Current Genomics* 12.5 (August 2011): 371–78.

20. Dennis Venema, "Genesis and the Genome: Genomics Evidence for Human-Ape Common Ancestry and Ancestral Hominid Population Sizes," *Perspectives on Science and Christian Faith* 62.3 (2010): 166–78.

21. Venema, "Genesis and the Genome," 175.

Evidence of Transitional Features in the Hominin Line

We discussed the concept of transitional features in chapter 4. Sometimes the term "transitional forms" is used, suggesting that a particular species is midway between two other species, popularly known as the "missing link." Evolutionists do not usually talk about transitional forms but instead talk about "transitional features" because it is the evolution of particular features over time that can be traced. But we can often find that a particular specimen has several transitional features. A number of the species discussed in the previous section have transitional features that illustrate a path from earlier mammals to modern humans.

One of the first transitional features in the fossil record is bipedal locomotion. Chimpanzees, gorillas, and other primates move around on two feet to some extent, but walking upright remains a human characteristic. We begin to find it in the early australopiths, who underwent a development in which the pelvis and the structure of the legs changed to facilitate walking upright. This did not happen all at once. Thus, although australopithecines seem to have evolved the ability to walk upright, they were still somewhat different from modern humans. Their pelvises were not the same, and some had prehensile big toes that would have helped in climbing but would have been a disadvantage in walking. Most primates were designed to live or at least spend part of their time in trees, for food, shelter, and protection. As well as their legs and feet, their arms and hands were suited to this mode of existence. The earliest bipedal creatures retained this ability to get around in the trees with their arms and hands. Thus, australopiths retained long arms and long curved fingers, indicating that they still spent some of their time in trees. Reduction in the length of the arms and changes in the shape of the hands to assist with the tasks of carrying, manipulation of tools, and technological creation paralleled over time the lengthening of the lower limbs. These changes continued toward more efficient upright walking.

Another transitional feature was dentition. The shape of the teeth and the chewing muscles are primarily determined by diet, but other factors come in, such as male display and fighting over mates. Male chimpanzees and some other apes have long canine teeth that are continually sharpened by contact with the anterior lower premolar. On the whole, hominins lack this enlarged canine and the mechanism for honing it (exceptionally, one early hominin, *Ardipithicus kadabba*, seems to have it). The canine tooth continues to shrink in later hominins, and the diastema (a gap between the teeth in chimpanzees and many other primates) generally disappears. The teeth and jaws as a whole become smaller (though, again, this is affected by the specific diet of the group in question).

Another set of transitional features relates to speech. The members of the genus *Homo* have "basicranial flexion," which means that the base of the skull arches up and determines the position of the larynx and other speech organs. The skulls of *Homo ergaster* exhibit the beginnings of a process that develops into a great basicranial flexion in modern humans. The hyoid bone in the neck with connections to the tongue and larynx and other speech organs is not usually preserved in fossil hominins. However, in later hominins it tends to be lower down. The larynx is high in the throat of apes and young humans. However, it is somewhat lower in *Homo ergaster*, which has an intermediate basicranial flexion, while its position in *Homo neanderthalensis* has been considerably debated. Yet many now think that *Homo neanderthalensis* did not differ significantly from *Homo sapiens*, in whom the larynx has moved farther down in the throat until it reached the position it occupies in modern humans.[22]

Speech also requires the ability to control the rib cage and breathing. An examination of the neural canals relating to the spine shows whether or not the individual had the fine control over breathing to produce articulate speech. In *Homo ergaster*, the neural canals imply only limited control, whereas in *Homo neanderthalensis* and *Homo sapiens*, they allow considerable control. Finally, there is brain size, which remained much the same with australopiths as with chimpanzees but gradually became larger with *Homo habilis*, more so with *Homo ergaster/erectus*, even bigger with *Homo heidelbergensis*, until it became its largest with *Homo neanderthalensis* and *Homo sapiens*. Some have speculated that in this process the brain also became restructured in *Homo sapiens* about fifty thousand years ago, though this is hard to demonstrate from fossil crania. There are many transitional features and forms in hominin evolution.

22. Klein, *Human Career*, 460, 650–53.

Interbreeding between Hominin Species

The genomes of *Homo neanderthalensis* and *Homo denisova* have now been mapped.[23] Modern descendants of African populations show no admixture of genes in common with *Homo neanderthalensis* and *Homo denisova*. However, Eurasians have between 1 and 4 percent of *Homo neanderthalensis* genes. One fossil of an early modern human from Romania had 6 to 9 percent *Homo neanderthalensis* genes, suggesting that an ancestor only four to six generations earlier had been a *Homo neanderthalensis*.[24] Surprisingly, it was found that peoples from Southeast Asia and Oceania (Melanesians, Australian Aboriginals, Polynesians, Fijians, and some inhabitants of the Philippines) had not only the same percentage of *Homo neanderthalensis* genes as Europeans but also between 4 and 6 percent of *Homo denisova* genes.[25] This and other studies indicate that interbreeding between early humans (*Homo sapiens, Homo neanderthalensis, Homo denisova*, and probably others) took place on some occasions. It is estimated that at least 40 percent of the *Homo neanderthalensis* genome has been preserved among the entire modern population, since *Homo neanderthalensis* genes in one person may not be the same as those in another.

We can conclude a number of things from this evidence. First, *Homo neanderthalensis* (and, by implication, *Homo denisova*) were human. Although *Homo neanderthalensis* (and their *Homo denisova* cousins) was a different species from *Homo sapiens*, they were closely enough related that they were able not only to mate but also to have offspring at least on some occasions.

23. A number of recent studies discuss the interbreeding (hybridization) of ancient humans, including David Reich et al., "Denisova Admixture and the First Modern Human Dispersals into Southeast Asia and Oceania," *The American Journal of Human Genetics* 89 (2011): 516–28; Morten Rasmussen et al., "An Aboriginal Australian Genome Reveals Separate Human Dispersals into Asia," *Science* 334 (2011): 94–98; Kay Prüfer et al., "The Complete Genome Sequence of a Neanderthal from the Altai Mountains," *Nature* 505 (January 2014): 43–49; Benjamin Vernot et al., "Excavating Neandertal and Denisovan DNA from the Genomes of Melanesian Individuals," *Science* (2016).

24. Qiaomei Fu et al., "An Early Modern Human from Romania with a Recent Neanderthal Ancestor," *Nature* 524 (2015): 216–19.

25. Sriram Sankararaman et al., "The Genomic Landscape of Neanderthal Ancestry in Present-Day Humans," *Nature* 507 (2014): 354–57; Svante Pääbo, "The Contribution of Ancient Hominin Genomes from Siberia to Our Understanding of Human Evolution," *Herald of the Russian Academy of Sciences* 65 (2015): 392–96.

Second, *Homo neanderthalensis* and *Homo denisova* were not just another variety of modern human; otherwise, the modern human genome would match the *Homo neanderthalensis* genome almost perfectly. We are dealing with a variety of humans; this fact is compatible with evolution from earlier primates. You cannot divide the primate group into the simple dichotomy "apes or humans." There is a spectrum of apelike creatures, a spectrum of humanlike creatures, and a variety of forms intermediate between the two. Third, humans have evidently kept and eliminated *Homo neanderthalensis* DNA by natural selection. Some *Homo neanderthalensis* DNA has apparently been helpful in the immune system. *Homo neanderthalensis* genetic inheritance of some Tibetan peoples seems to have benefited them in adjusting to living at high altitudes. However, some parts of the modern human genome have no *Homo neanderthalensis* DNA; one of these is in the area of a gene relating to speech. This might suggest that *Homo sapiens* genes were better suited for speech than *Homo neanderthalensis* genes. Finally, sub-Saharan Africans have no "foreign" DNA. Europeans and some people from Asia have an admixture of Neanderthal genes. This suggests that the original population moved out of Africa into Europe and parts of Asia where they encountered *Homo neanderthalensis*. The Melanesians and other peoples in Southeast Asia possess a certain percentage of *Homo neanderthalensis* and *Homo denisova* DNA. They evidently migrated through a *Homo neanderthalensis* area (perhaps Eastern Europe), but also encountered *Homo denisova*, probably not in Central Asia but possibly in Southeast Asia itself.

Conclusions

Fossil discoveries and DNA research continue to inform our understanding of human origins. The evidence demonstrates that all the primates are closely related. All early hominins are genetically connected. The fossil record documents a variety of transitional features from ape to human over a long period, with various creatures in transitional positions. Although much remains to be learned, there seems to be no doubt that modern humans evolved from earlier primates over millions of years.

9

THE ADAM DEBATE

An Iconic Human

The previous chapter examined the fossils of humans and humanlike creatures. How do fossils of primitive humans inform interpretations of the first chapters of Genesis and, in particular, Adam and Eve?

Although some religious people would deny the scientific data (or at least the scientific interpretation), many believers now accept the fossil record as evidence for human evolution.[1] Does accepting evolution create a conflict with faith?

Some Christians, especially among the ranks of evangelical Protestants, are on "the brink of crisis" with regard to Adam and Eve.[2] A number of signs point to the threat of such a crisis. One such indication is the recent book, *Four Views on the Historical Adam*. The introduction by the editors Matthew Barrett and Ardel Caneday offers a careful and nuanced discussion.[3] The widely circulated evangelical publication *Christianity Today* focused on the

1. See the discussions in, among others, J. P. Moreland and John Mark Reynolds, eds., *Three Views on Creation and Evolution* (Grand Rapids: Zondervan, 1999); Thomas B. Fowler and Daniel Kuebler, *The Evolution Controversy: A Survey of Competing Theories* (Grand Rapids: Baker Academic, 2007); Scot McKnight and Dennis Venema, *Adam and the Genome: Reading Scripture after Genetic Science* (Grand Rapids: Brazos, 2017); William T. Cavanaugh and James K. A. Smith, eds., *Evolution and the Fall* (Grand Rapids: Eerdmans, 2017). Note also the discussion in Christopher Lilley and Daniel Pedersen, eds., *Human Origins and the Image of God: Essays in Honor of J. Wentzel van Huyssteen* (Grand Rapids: Eerdmans, 2017).

2. This statement is borrowed from John R. Schneider, "Recent Genetic Science and Christian Theology on Human Origins: An 'Aesthetic Supralapsarianism,'" *Perspectives on Science and Christian Faith* 62.3 (2010): 199.

3. Matthew Barrett and Ardel B. Caneday, eds., *Four Views on the Historical Adam*, Counterpoints: Bible and Theology (Grand Rapids: Zondervan, 2013).

search for the historical Adam in June 2011. The article delineated a series of opinions and positions about the historicity of Adam.

The journal of the American Scientific Affiliation, *Perspectives on Science and the Christian Faith*, devoted the September 2010 issue to the theme "Reading Genesis: The Historicity of Adam and Eve, Genomics, and Evolutionary Science," with a range of opinions expressed. As a final example, in August 2015 the "Books and Culture" feature on the BioLogos website hosted an online symposium on Adam and Eve. Chaired by John Wilson, the symposium featured original articles and follow-up responses by Peter Enns, Karl Giberson, Denis O. Lamoureux, Hans Madueme, "Hal" Lee Poe, John Schneider, William VanDoodewaard, and John H. Walton. The positions ranged from those who argued that a historical Adam and Eve were essential to the doctrines of original sin and redemption in Christ to those who felt that a literary or symbolic Garden of Eden story was deeply meaningful and fully sufficient.

The Creation of the Humans (Genesis 1:26–28)

The first account of the creation of humans is part of the creation account of the world. In Genesis 1:26–28 God says,

> "Let us make man in our image (*ṭselem*) according to our likeness (*demut*), and they shall rule over the fish of the sea and the birds of heaven and the (land) animals and all the earth and all swarming things that swarm on the earth." And God created the man in his image, in God's image God created him, male and female he created them. And God blessed them and God said to them, "Be fruitful and multiply and fill the earth and subdue it, and rule over the fish of the sea and the birds of heaven and over all living things swarming upon the earth."

I have quoted this at length to show how beautiful the language of the passage is. In its simplicity and its repetition, it borders on the poetic. In spite of the simple language, there is a great deal being described.

The first thing to notice is that humans are created in God's image (*ṭselem*). What does this mean? It is often used of a physical likeness, such as a statue or drawing. Is it referring to God's shape and appearance? We tend to depict God in human shape, but why? Some of the ancient Greeks debated the shape of divine beings. The pre-Socratic philosopher Xenophanes said that if horses and cattle could conceive of gods, they would be in the shape of horses and cattle.[4] Humans are male and female. Would God be in the shape of a male or a female? After all, even though God is often referred to in Hebrew in the masculine gender, there are also female references to God (as a mother; as a woman [Deut 32:9–14; Isa 66:13]). God is also presented as a bird (Ps 91:4; Deut 32:11; cf. Matt 23:37). Should we think of God as an eagle or a giant chicken? The point is that God is spirit, and we should not think of human shape or any other shape as "God's image," even though he has sometimes manifested himself in human form.

Some have suggested that God's image often relates to the king in ancient Near Eastern texts. However, Genesis 1:28 is clearly referring to all humanity, not just kings and rulers. The one thing we think of as particularly human is the mind (or intellect) and all that goes along with it: culture, intellectual pursuits, philosophy, arts, design, technology, and so on. The mind of human beings differs from those of the rest of living things, at least in degree if not in kind.

Another point in the narrative is that humans are male and female. There is no distinction made between them – they are both creations of God. Indeed, they are both "man" in the sense of being human. The Hebrew word "man" can designate both a (male) man and a human (whether male or female). There is no suggestion of the female being taken out of or in any way being derived from the male in Genesis 1:26–28.

Adam and Eve in the Garden of Eden (Genesis 2:4–3:24)

How are we to understand Genesis 2:4–3:24 in relation to Genesis 1:1–2:3? For almost two centuries, biblical scholars treated the two stories of creation as two separate accounts of creation. Is this only scholarly insolence, an attempt to complicate what is simple? After all, many readers have seen

4. See Clement of Alexandria, *Miscellanies.* 5.110; 7.22.

chapters 2–3 as simply the detailed account of the creation of humans in Genesis 1:26–28. Yet a careful reading reveals some anomalies.

Genesis 2:4 begins, "These are the generations (*toledot*) of the heavens and the earth in their being created." This might sound like a heading, since elsewhere in Genesis *toledot* has an organizing function, but a number of researchers have argued that it usually ends a section (6:9; 10:1; 11:10, 27; 25:12, 19; 36:1, 9; 37:2). This is why many begin the story with the second half of Genesis 2:4: "In the day when YHWH God made earth and heaven," continuing in verses 5–7, "before any shrub of the field was on the earth and before any herb of the field had sprouted, because YHWH God had not caused it to rain on the earth, and there was no man (*'adam*) to work the soil. Then a water stream [precise meaning of the Hebrew word uncertain] went up from the earth and watered the whole face of the soil. Then YHWH God created the man (*ha'adam*) from dust of the soil and breathed into his nostrils the breath of life, and the man became a living soul (*nefesh*)."

As the narrative begins, there are no plants on the earth. This is because it has not rained and because there are no humans to till the earth. God sends a water source of some sort over the earth to water it, as he creates the first man. This is completely different from Genesis 1 in which the plants were created on the third day – with no problems about rain or water – and humans were not created until day six. As the story unfolds, Genesis 2–3 looks like another creation story, one in some ways parallel to Genesis 1:26–28 but in other ways rather different. This will become important as we consider the significance of the story of Adam and Eve.

As the story continues, we find the man referred to sometimes as *'adam*, which can be "man" or can be the name "Adam" (2:20); most of the time, though, the word has the article, which translates as "the man" (2:8, 15, 16, 18, 19, 20, 21, 22, 23, 25; 3:8, 9, 12), even though some translations still at times render it as "Adam." Whereas in Genesis 1:26 humans were created male and female on the sixth day, Genesis 2 has the man alone created. He then names all the animals (birds and land animals, though apparently not the fish). Yet within all these birds and animals, "for Adam, there was not found any helper suited to him" (2:20). This is very strange. In Genesis 1, various animals, birds, fish, and other living things were created and told to multiply, suggesting that both sexes were created together for most of them. Yet strangely, the man of Genesis 2 has no female counterpart. A suitable helpmate is sought among all the birds and animals. None is found.

Can we believe that God was seriously looking for a counterpart to the female among the various birds and animals that Adam was naming? If we try to read the story literally, it appears strange indeed. But if we read it symbolically, it makes perfect sense. One might notice also that Adam is placed in the Garden of Eden twice, once in 2:8 and then a few verses later in 2:15.

In Genesis 2:21–24 we have the creation of Eve. God took one of his *tselah*. This is traditionally translated "rib" in most English versions. The Hebrew word can mean "rib" but also "side" (including an architectural reference). One commentator has recently argued that "side" is the correct translation, and that Adam was essentially split in half, with half of him used to create Eve.[5] This meaning is possible. Others think that Eve was created from Adam's genitals.[6]

Genesis 3 is the story of the serpent and the tree of knowledge of good and evil. This chapter is often thought to describe the "fall of man" (though Jewish tradition does not know of a "fall of man"). The fall of man is an important Christian doctrine, as we shall see when we consider the apostle Paul's teaching. At its heart is the sin of Adam and Eve, not some sort of change of nature as is sometimes thought. This is very clear when we read the passage in context and do not import later interpretations.

First, there is the overall message of Genesis 2–3: the man and woman are commanded not to eat of the fruit of the tree in the middle of the garden: "From every tree of the Garden you may certainly eat, but from the Tree of the Knowledge of Good and Evil you may not eat of it, for in the day [or "when"] you eat from it you shall certainly die" (2:16–17). Adam and Eve did eat from the tree, and they did die. But what changed? Did they "fall"? No, nothing changed, because they were created mortal and would eventually have died anyway. The implication of Genesis 3:22–24 is that humans might have had the opportunity for eternal life, but the exact route to this eternal life is never spelled out. The impression is that there may have been

5. John H. Walton, *The Lost World of Adam and Eve: Genesis 2–3 and the Human Origins Debate* (Downers Grove: InterVarsity Press Academic, 2015), 77–79.

6. Scott F. Gilbert and Ziony Zevit, "Congenital Human Baculum Deficiency: The Generative Bone of Genesis 2:21–23," *American Journal of Medical Genetics* 101.3 (2001): 284–85; Ziony Zevit, "Was Eve Made from Adam's Rib — or His Baculum?" *BAR* 41.5 (2015): 32–35.

an opportunity to obtain eternal life, which was lost, but no explanation is given. As we shall see, the question of eternal life is discussed in the New Testament, making use of the Garden of Eden episode, but the story in the Hebrew Bible comes to an abrupt end, with humans subject to mortality. Yet, to emphasize my point, Adam and Eve were created mortal; they did not "fall" into mortality.

In 3:1 the serpent is introduced. It is normally assumed that the serpent is Satan or the devil. Yet the serpent is never identified as such in Genesis. On the contrary, the serpent is simply subtler than any of the other *animals* (Genesis 3:1). Although the serpent in the Garden of Eden eventually becomes identified with the devil, this is nowhere found in the Hebrew Bible, nor does Paul say that the serpent is anything other than a cunning animal (1 Cor 11:3). It is true that Satan, the devil, is referred to as "the old serpent" in Revelation 12:9; 20:2, but no mention is made of the Garden of Eden. What should we make of a talking serpent who aims to deceive Eve and make her sin? This is a morality tale – a theological narrative, like the parables – not a historical account. Interestingly, his arguments to Eve are the very arguments that one would make with oneself if debating internally whether to disobey God! They constitute the sort of arguments that any parent hears from a child who has done something he or she was told not to do.

But there is a further indication that the serpent is first and foremost an animal, found in Genesis 3:1: "And the serpent was the cleverest of all the animals of the field *that YHWH God had made*." Note the last phrase (in italics): the serpent is one of the creatures made by God, along with the other animals. This is not a great supernatural opponent of God, an angelic deceiver, but a simple animal among the many animals. Apart from being clever, this animal differs from the other animals in only one major aspect: it talks. But this is a sign that we are reading a story with a theological message, not a description of human origins.

We have another indication in the curses that God pronounces when the disobedience of Adam and Eve is discovered. The curse on the serpent is simply a description of its mode of life – it lives by crawling on its belly, and there seems to be a natural human antipathy toward it (though this antipathy is not confined to snakes but also applies to other "creepy crawlers" such as spiders). But then the curse on the woman is very strange: "I shall make great your birth pains; in pain you shall bear children." All mammals have a form of labor in order to give birth, since the offspring has to be expelled from the womb. Humans differ from other animals in that the baby has a very large head because of development of the brain.

Ideally, the baby would be brought to full development before birth, as is the case with most mammals, but with humans this is not possible because the head would be too large for any birth canal. Human babies are born underdeveloped, which is why they are so helpless in the first year compared to many other animals. Thus, human birth is a compromise between letting the child mature as far as possible while still being small enough to pass through the mother's pelvis. The female pelvis can only grow to a certain size, which means that the baby's head must not be too large to pass through the birth canal. Therefore, human birth can be subject to greater trauma than is the case with many animals. Pain in childbirth is not a curse because of Eve's sin but is the natural result of this evolutionary compromise between the limits on the size of the pelvis and the maximum development of the baby at birth. The description of the human condition in Genesis 3:16 has nothing to do with the sin of a first mother.

Finally, in Genesis 3:17–19 we have God's pronouncement on Adam: "Cursed is the ground for your sake. In pain you shall eat all the days of your life. Thorns and thistles shall sprout before you and you shall eat the herb of the field; in the sweat of your face you shall eat bread until you return to the earth, because from it you were taken, for dust you are and to dust you shall return." Were thorns and thistles created at this time, or were they not part of the creation of plants from the start? Thorns and thistles are a part of nature. When farmers cultivate the soil, any seeds around will sprout, but in order to make sure that only the right crop grows, they plow or hoe or otherwise remove plants they do not want to grow there. There is nothing sinister about these other plants; they just happen to be out of place from the farmer's point of view. Growing crops requires a good deal of work; farming is a labor-intensive occupation. Anyone who has grown up on a farm knows how hard the work can be, but it is the nature of the occupation. The plight of the farmer is well described here, but there was no physical change of plants – no new plants suddenly springing up – or of life as a farmer at this time.

So what we see in Genesis 3:14–19 is a description of the world in which humans (and the serpent) lived. Snakes crawl; they bite humans or are killed by humans. But this is the nature of a snake's life. No suggestion is made that snakes had legs and lost them at this point. Women experience labor to give birth to children. For most, it is not a pleasant experience (until the end when they experience the joy of having a new healthy baby). Most people lived by agriculture at this time, requiring a lot of hard work to produce enough food for themselves and their family. Adam and Eve sinned and were punished. But nothing is said *in Genesis* about a fall. Nothing is said here about a change in nature. The "curses" are really just a description of life as it was lived.

An abundance of details cry out to the reader that the story is not meant to be taken literally. There are clear indications that this is *another* story about the creation of humans, but it is different from Genesis 1:26–28. Adam is placed in the Garden twice (2:8, 15). In it Adam strangely considers taking a mate from among the various animals, but none suitable is found. Even though male and female seems to have been taken for granted with the rest of mammal creation, Adam has no female counterpart: she has to be created especially for him, almost as an afterthought. One of God's creatures suddenly starts talking to Eve, out of the blue. The man and woman are given mortality as punishment for their sin, even though they were already very mortal. Then there are the curses: the serpent is told he must crawl on the ground, which he had already been doing for some time; the woman has to experience labor to produce children, which is the case with all mammals; and farmers have to work the land and weed out the natural growth of plants, because crops do not sow and cultivate themselves.

Christianity would develop a doctrine of the fall of man and make use of the story about Adam and Eve to explain it. Genesis 2–3 acts as a sort of allegory to illustrate the consequences of sin and the human condition, as the New Testament passages make clear. Much more could be said about Genesis 2–3, but my aim has not been to write a commentary on these chapters but rather to discuss their implications. We are now ready to look at some of the relevant New Testament passages.

Adam and Eve according to the Apostle Paul

Adam and Eve appear in several passages in the New Testament, but I want to focus on passages among the letters of Paul. The first is Romans 5:14. In the wider context, Paul is making a point about sin (Rom 5:12–21). Sin and death entered into the world through one man, though all humans are subject to death *because all have sinned*. But just as death came about by one person, so life will come through one person: grace, righteousness, and life come through Jesus Christ. Paul also mentions that "Adam was a type of the one to come" (5:14).

This reference to "type" is interesting because it refers to the exegetical technique of allegory. Allegory is found sporadically in the Hebrew Bible, early Judaism, and the New Testament. However, it was a mode of exposition especially favored by certain Jewish commentators, such as Philo of Alexandria.[7] We also find some examples in Paul. For example, in Galatians 4:21–31 he gives an allegory based on Sarah and Hagar, from the book of Genesis. Abraham had two sons, one born of a slave and one born of a free woman. These represent two covenants, but these also represent two Jerusalems, the earthly Jerusalem and the heavenly. Hagar also represents Mount Sinai; that is, the Mosaic law, though no counterpart is given for this aspect of the allegory. The important thing is that the Galatian Christ-believers are children of the free woman, the heavenly Jerusalem.

Now, we can ask, is Paul trying to tell us that there was a literal, historical Isaac and a literal, historical Ishmael? Is this his purpose? Of course not. He is using figures from the book of Genesis to make a point. He does not even mention Ishmael by name. Whether the figures mentioned in the allegory are historical or only literary is not relevant: the important thing is what they represent. Would it make any difference to the theological message if Abraham, Isaac, Ishmael, Sarah, and Hagar were simply figures in a fictional story? Not at all.

Let us consider another passage in which Paul uses a passage from the Hebrew Bible to make a theological point: 1 Corinthians 9:8–12. He is ar-

7. See Lester L. Grabbe, *Etymology in Early Jewish Interpretation: The Hebrew Names in Philo*, BJS 115 (Atlanta: Scholars Press, 1988). I explain Philo's method and give extensive bibliography on the use of allegory in the ancient world.

guing that he has a right to be free to live his life, which includes making a living from preaching the gospel. He then quotes Deuteronomy 25:4, "You shall not muzzle an ox while it is threshing (grain)." What was the purpose of this law? One might think of more than one intent, but surely one purpose was concern for the welfare of the animal doing the work. Yet Paul is not interested in that, and even states, "Is it about oxen that God is concerned, or does he not speak entirely with regard to us? For he writes concerning us that the plowman should plow in hope and the reaper should reap in hope" (1 Cor 9:9–10). Is God concerned about the welfare of animals? Naturally! Yet Paul brushes past that to make the point that he should have his physical needs provided for as he takes the message of the gospel to others. The important issue for him is not the original context and meaning but his application of the passage to his contemporary situation.

We can now return to Romans 5:12–21. Paul's aim is not to declare whether or not there was a literal Adam and Eve. He is using Adam as a counterpart to Jesus, illustrating how sin and death entered the world through Adam – though we are all subject to death because we have all sinned – but we can gain grace and life through the "second Adam" who is Jesus. Adam did not lose immortality; he was *never* immortal. He was mortal from his creation. The text makes it clear that he was prevented from gaining immortality.

This same theme is continued in 1 Corinthians 15:20–28. Again, Paul is contrasting death through Adam and life through Jesus: "For as death (came) through a man, so also resurrection of the dead (came) through a man. For as in Adam all died, so also in Christ shall all be made alive" (vv. 21–22). Paul continues his discussion of the resurrection, but says some verses later, "Thus also it has been written, 'The first man Adam became a living soul,' the last Adam a life-giving spirit. . . . The first man (was) from the earth, the second man from heaven" (vv. 45–47). Paul is interested in expounding the resurrection. Death entered the world through the first Adam, but resurrection and life through the second Adam. Paul is using a figure of speech, creating an image, and illustrating a point of theology. He has no interest in discussing science, nor should we attempt to extract such information from these passages.

Should we assume that Paul had modern scientific knowledge? Must we assume that he knew the earth was round and that it went around the sun? Did he know about thousands of galaxies? Did he know about the solar system? Was he aware of the big bang theory? Does it make any difference to his teachings if he did – or did not – have such knowledge?

Why should we assume that Paul had a modern knowledge of science? Although some Greek scientists had theorized that the earth went around the sun, the standard model was the Ptolemaic system in which the earth was the center around which the sun, the moon, and planets circulated. It is possible, of course, that Paul believed in the heliocentric theory of the universe, but the chances are that he accepted some version of the Ptolemaic theory, to the extent that he had knowledge of astronomy. Let us assume that he accepted a geocentric model of the universe: Would that have made an iota of difference to his teachings? Should it make us revise our view of the world and the universe? The issue is not Paul's knowledge of science but the theological message he is communicating. To him Adam is a symbolic figure who serves as a counterpart to Jesus. Adam represents mortality and death, just as Jesus represents immortality and life.

Some may say, "How can Paul use Adam as a symbol if he did not actually exist? Jesus had to exist. Why not Adam?" True, Jesus had to exist because of what he did and what was done through him. However, symbols do not necessarily have to exist in the real world to have meaning. We can use Adam as a symbol of the first man, of sin entering the world through him, because we have all sinned and are all mortal. This is how Jesus and Adam differ: only Jesus could make the sacrifice he did and do the things he did, but Adam represents us all we have all sinned. We can accept the profound theological symbolism without worrying about whether paleontologists have proved whether or not there was an Adam.

We can go back to Paul and take as an example the Eucharist (called "the Lord's Supper" by many Christians), which he mentions in 1 Corinthians 11:23–25. This is based on Jesus's last meal with his disciples (Mark 14:22–24; Matt 26:26–28; Luke 22:19–20). However, the most graphic description is given in John 6:53–56:

> Then Jesus said to them, "Truly, truly, I say to you, if you do not eat the flesh of the son of man and drink his blood, you do not have life in yourselves. The one eating my flesh and drinking my blood has eternal

> life, and I shall raise him up in the last day. For my flesh is truly food, and my blood is truly drink. The one eating my flesh and drinking my blood remains in me and I in him."

Does the bread and wine have to be the actual flesh and blood of Jesus? This is the way it was interpreted for centuries. The doctrine of transubstantiation, as it was called, was clearly a belief of the medieval church. Protestants thought differently: they interpreted the bread and wine as a symbol of Jesus's flesh and blood, even though it represents eternal life and all that entails. They are not *literally* the flesh and blood of Jesus; they are only *symbolically* his flesh and blood. They are a figure that represents the blood that he shed and the sacrifice of his body that he made. But if eating the flesh and drinking the blood of Jesus can be interpreted symbolically, even though they are essential for the Christian life, why cannot the sin of Adam – and his existence – be taken as symbolic? We do not literally eat the flesh and drink the blood of Jesus. Why should Adam have literally sinned for his example to be meaningful?

Adam and Eve in the Light of Theology

We have considered some biblical passages relating to Adam and Eve. I have expressed my understanding of these passages. There will naturally be differences of opinion. Yet the accumulated scientific data limits interpretations of the biblical passages.

Many people of faith recognize the validity of science about the development of living things. Yet, some still feel a twinge of uncertainty when it comes to human development. Are humans really the result of evolution from single-celled creatures until finally a branch of the primates developed into humans? The scientific evidence is overwhelming. How can this picture be reconciled with the biblical picture of Adam and Eve and the Garden of Eden?

As discussed above, many believing scientists have no problem reconciling evolution with their religious faith, whether of the Bible or otherwise. However, I expect some readers – especially evangelical Christians – may be troubled by the question of how to reconcile the stories about Adam and Eve with the emerging science of human evolution.

In what follows, I want to concentrate on showing how some important evangelical scientists, theologians, and biblical scholars have reconciled the references to Adam and Eve with the scientific evidence. Religious believers will not always have the same interpretation of the passages in question, but it may be helpful to see how others with similar religious views have dealt with the issues.

First, Denis O. Lamoureux, as noted earlier, gave up Christianity and became an atheist as a young man. However, after a conversion back to Christianity, he embraced young-earth creationism and completed a doctorate in theology. When challenged on evolution, he determined to pursue a degree in biology and devote himself to refuting Darwinism. His work with fossils convinced him that evolution was correct and young-earth creationism was wrong, but he remains a strongly committed evangelical Christian advocating "evolutionary creation" (usually classified as a form of theistic evolution).[8]

Lamoureux opposes "scientific concordism," which he defines as "the assumption that God revealed scientific facts to the biblical writers thousands of years before their discovery by modern scientists."[9] In his view, scientific concordism fails because the Bible is not a book of science. Further, the science in the Bible is ancient science. For Lamoureux, Adam is not the tail end of evolution. Rather, Adam's story is our story: "To conclude, I do not believe that there ever was a historical Adam. Yet he plays a pivotal role in Holy Scripture. Adam functions as the archetype of every man and woman. In Genesis 2 and 3, he is an incidental ancient vessel that delivers numerous inerrant spiritual truths. His story reveals that the Creator has set limits on human freedom. We are accountable before God, and a failure to obey His commands results in divine judgment."[10]

John H. Walton, professor of Old Testament at Wheaton College and Graduate School, has written extensively on the early chapters of Genesis, including his PhD thesis on the Tower of Babel at Hebrew Union College.[11]

8. See Lamoureux, *Evolutionary Creation*. Also, Denis O. Lamoureux, "No Historical Adam: Evolutionary Creation View," in *Four Views on the Historical Adam*, ed. Matthew Barrett and Ardel B. Caneday (Grand Rapids: Zondervan, 2013), 37–88.

9. Lamoureux, *Evolutionary Creation*, xv, 16–18; Lamoureux, "No Historical Adam," 45.

10. Lamoureux, "No Historical Adam," 65.

11. See, among others: John H. Walton, *Genesis 1 as Ancient Cosmology* (Winona Lake, IN:

His book specifically on Adam and Eve is set out as a series of propositions, which he then proceeds to explain and justify. He accepts that Adam and Eve were real individuals but not necessarily the first or only human beings. Indeed, although he does not go into very much detail, he appears to accept the common evolutionary view that modern humans are descended from a small population of people who lived about one hundred and fifty thousand or so years ago. Adam and Eve were one couple from this group whom God chose for a special reason.[12]

However, Walton's concern is not to defend the actual existence of Adam and Eve but to show their significance in the story of Genesis 2–3. Adam and Eve serve as "archetypes"; that is, they represent all of us, with the same human characteristics that we all possess. They have the same sinful human nature.[13] For Walton, Adam's creation from dust and the creation of Eve from his "rib" are not statements of material origin but affirmations of our nature. The New Testament is not interested in Adam and Eve as the biological progenitors of other humans but in their function as archetypes who represent mortal humanity.

N. T. Wright is the former bishop of Durham and a leading light in the evangelical wing of the Church of England. Since his retirement as bishop, he has become research professor of New Testament and Early Christianity at St. Mary's College in the University of St. Andrews in Scotland. He contributed a section for John Walton's book, *The Lost World of Adam and Eve*.[14] His concern in this excursus is to discuss the apostle Paul's use of Adam in various passages. Wright argues that the science of cosmology and human origins has become "muddled up" with the soteriological question of how sinful humans are to be saved. Paul's concern is the kingdom of God, and he uses Adam to make his point. Adam has a similar function or vocation as Israel: to carry through God's project for the whole creation, for God's redeemed people to take back control of a new creation as royal priests of

Eisenbrauns, 2011); John H. Walton, *The Lost World of Genesis One: Ancient Cosmology and the Origins Debate* (Downers Grove: InterVarsity Press Academic, 2009); Walton, *The Lost World of Adam and Eve*.

12. Walton, *The Lost World of Adam and Eve*, 177–78.

13. See especially Walton, *The Lost World of Adam and Eve*, 70–95.

14. N. T. Wright, "Excursus on Paul's Use of Adam," in Walton, *The Lost World of Adam and Eve*, 170–80.

the creator. The story of Adam of Eve is the story of how humans messed up their responsibility of steering creation.

Francis Collins, as noted earlier, is one of the best-known scientists of faith, because of the publicity given to him for his work in genetics, particularly the first cataloging of the human genome by the Human Genome Project under his leadership. He is currently director of the National Institutes of Health (Bethesda, Maryland). He has received the Presidential Medal of Freedom, the National Medal of Science, and appointment to the Pontifical Academy of Sciences (by Pope Benedict XVI). While studying for his doctorate, he moved from a youthful agnosticism to atheism. After completing a PhD in physical chemistry, he went on to earn an MD. It was while practicing medicine that he was forced into examining the question of belief and nonbelief, which resulted in his accepting belief in God and embracing the Christian faith.[15]

Collins is a prominent scientist. His work on DNA has confirmed the correctness of Darwinian evolution. He regards himself as a theistic evolutionist, though he found problems with the term, preferring to use "Biologos" to describe his position. He founded the BioLogos Foundation in 2007 to contribute to the discussion on the relationship between science and religion and also to argue for the compatibility between science and the Christian faith.

> As noted previously, studies of human variation, together with the fossil record, all point to an origin of modern humans approximately a hundred thousand years ago, most likely in East Africa. Genetic analyses suggest that approximately ten thousand ancestors gave rise to the entire population of 6 billion humans on the planet. How, then, does one blend these scientific observations with the story of Adam and Eve? . . . Many sacred texts do indeed carry the clear marks of eyewitness history, and as believers we must hold fast to those truths. Others, such as the stories of Job and Jonah, and of Adam and Eve, frankly do not carry the same historical ring. . . . I do not believe that the God who created all the universe, and who communes with His people through prayer and spiritual

15. See Collins, *The Language of God.*

insight, would expect us to deny the obvious truths of the natural world that science has revealed to us, in order to prove our love for him.[16]

Denis Alexander is emeritus director of the Faraday Institute for Science and Religion at St. Edmund's College, Cambridge. In his book *Creation or Evolution: Do We Have to Choose?*, he outlined several models for Adam and Eve, some of which have already been discussed. He favors what he calls the *Homo divinus* Model.[17] This model accepts the results of paleoanthropologists, archaeologists, and ancient Near Eastern scholars. He argues that God chose a couple of Neolithic farmers in the ancient Near East to manifest himself in a more direct way. They were not the first or only humans, since people had a long history, as shown by the study of primate evolution. But these two individuals were the first humans who were "truly spiritually alive in fellowship with God." They provided the spiritual roots of the Jewish faith and formed the basis of the story of Adam and Eve. Therefore, he believes that Adam and Eve were real figures in history, but in a somewhat different way from the story in Genesis 2–3.

Peter Enns is currently Abram S. Clemens Professor of Biblical Studies at Eastern University in St. Davids, Pennsylvania. In his book on Adam, he concludes with nine theses on how he thinks Adam is to be understood.[18] He makes the point that the scientific model and the biblical model that assumes a literal reading are incompatible because they "do not speak the same language." He sums up the situation in a robust manner: "One cannot read Genesis literally – meaning as a literally accurate description of physical, historical reality – in view of the state of scientific knowledge today and our knowledge of ancient Near Eastern stories of origins. Those who read Genesis literally must either ignore evidence completely or present alternate 'theories' in order to maintain spiritual stability. Unfortunately, advocates of alternate scientific theories sometimes keep themselves free of the burden of tainted peer review. Such professional isolation can encour-

16. Collins, *The Language of God*, 207–10.

17. On the various models, see *Creation or Evolution: Do We Have to Choose?* (Oxford: Monarch Books, 2014).

18. Peter Enns, *The Evolution of Adam: What the Bible Does and Doesn't Say about Human Origins* (Grand Rapids: Brazos, 2012).

age casually sweeping aside generations and even centuries of accumulated knowledge."[19]

Although he accepts the possibility that "God took two hominid representatives" and began the human story with them, this is not the biblical story but our own creation.[20] He is happier with the Adam story as simply an ancient story with a meaning. It is the story of the failure to fear God and attain wise maturity. According to Enns, the apostle Paul read the Adam account in light of his experience of the risen Christ: "One can believe that Paul is correct theologically and historically about the problem of sin and death and the solution that God provides in Christ without also needing to believe that his assumptions about human origins are accurate. The need for a savior does not require a historical Adam."[21]

His final point is very significant: one cannot simply graft evolution onto an evangelical Christian faith. Rather, a "true rapprochement between evolution and Christianity [or perhaps any other religion] requires a synthesis."[22] We have to accept the findings of science. Although scientific discoveries are always provisional and may be revised in the light of new data, we cannot deny basic scientific facts, nor can we ignore the findings of scholars of the ancient Near East in relation to the biblical world. Our theology must grow and develop in the light of this new knowledge.

Conclusions

Along with Lamoureux, Enns, Brown, and others, I have argued that Adam and Eve were characters in a theological narrative. They did not have to exist in order to have great meaning for Christians, Jews, and other religious people. As figures in a story, Adam and Eve provide rich and meaningful symbolism for important theological and religious truths. The fact that they are symbols rather than real people does not negate their spiritual reality. Also, it seems clear to me that it is very difficult to fit a literal Adam and Eve into the actual scientific data that we have on human origins and development.

19. Enns, *The Evolution of Adam*, 137.
20. Enns, *The Evolution of Adam*, 139.
21. Enns, *The Evolution of Adam*, 143.
22. Enns, *The Evolution of Adam*, 147.

However, there are other models. Others have affirmed the actual existence of Adam and Eve while still accepting the scientific explanation of human origins. I think there are problems with this position, as already noted. In the end, it seems to me that there is not much difference in our overall stance. We share an acceptance of scientific discoveries and explanations. We think that the Bible has been wrongly interpreted by many sincere believers who are not fully aware of how it was written and how we received its words and its message.

Those who believe Adam and Eve existed but were not the first or only humans have ultimately given a symbolic interpretation to the Genesis stories and have made the theological archetypical meanings the prime concern. Affirming that Adam and Eve were actual persons does not (in my view and the view of many others) change the emphasis or the ultimate significance of the biblical message. We are mortal and sinful individuals not because of two human beings in the distant past but because of who we are and the nature with which we were born. That nature was not put there by anything Adam did but because it is part and parcel of being human. The story of Adam and Eve makes that clear; we do not have to historicize the story for the fact to be obvious and for the need for redemption to be apparent. Whatever one's religion, human nature is the same. Each religion seems to recognize the need to transcend our essential nature to be something different and better and to acquire a spiritual life that we do not naturally possess.

The science of cosmology and human origins continues to advance. Coming to terms with this new knowledge demands new interpretations of the Bible. Theological understanding cannot remain static. Just as scientific theories change with new data, so too theology and biblical interpretation must change to accommodate new knowledge of the world.

10

"IN OUR IMAGE"

Reflections of a Biblical Scholar

For most Christians, Jews, or Muslims, the question of the origin of the universe has one answer: God. Life, in all its majestic complexity, owes existence to the one source of all: God, the creator of things seen and unseen. Many may be surprised to learn that no less than forty percent of scientists express a belief in a personal God, and even more acknowledge the idea of God even if they do not understand this notion in personal terms. The idea that science drives one away from faith is a common assumption but one not substantiated by the facts. While it is true that some leading scientists have embraced atheism because of their scientific work, other scientists who began as atheists have embraced a personal faith at least in part due to their experiences as scientists.

Some recent scientific studies have emphasized the extent to which we live in a world that uniquely possesses the qualities that make life possible. According to all known natural laws, there are certain parameters outside which no life is possible. It takes a planet whose physical attributes fit a rather narrow range in terms of temperature variation, electro-magnetic radiation, and liquid water (to name some of the key characteristics) for any sort of life to exist, even bacteria. For higher life-forms, the range of physical constraints is even narrower. This optimum range has recently been dubbed the "Goldilocks effect."[1] That our world is so ideal to life seems to be a "divine design" according to some Christians and people of other faiths.

1. Paul Davies, *The Goldilocks Enigma: Why Is the Universe Just Right for Life?* (London: Allen Lane, 2006).

That the earth is in this "Goldilocks" ideal does not mean that it is the only possible site for life. The vastness of the universe would suggest that other planets must exist in other solar systems that also have sufficiently ideal physical properties to support life. However, insofar as our knowledge of the universe is limited, we know only that earth is the sole planetary body that is known to support life. So far as we currently know, it is unique in the vast universe of trillions of stars and their planets.

Regardless of whether there is intelligent life only on earth or elsewhere in the universe, we can affirm an important aspect of the wonder of the universe that goes beyond merely its beauty and its overwhelming span of size, distance, and greatness. The structures within the universe, from the largest interstellar bodies to the microscopic and even atomic worlds, are also causes for wonder. Design is an intrinsic part of all that we see. Integrated systems are found in astronomy, geology, and the realm of living things. Our own bodies – the bodies of every living creature – are made up of intermeshed, coordinated, and integrated systems that make life possible. It evokes a state of wonder; we perceive divine creation.

We can, therefore, affirm with all people of faith that this universe is indeed the result of God's creation. Most people of faith readily assert this with conviction. However, that affirmation does not erase the concerns of this book. While we can and do affirm that God created the heavens and the earth and all life within them, the question of *how* remains inescapable. Did he, over the course of seven days, bring everything into existence by a series of divine commands, as outlined in Genesis 1? Or did he use some other mechanism? What was the mode of divine creation of living things?

Most people of faith would agree on what God *could have done*. The question is: What *did* he do? If we accept the notion that every "kind" of animal is a special creation, we would expect the designed plant or animal to be perfectly tailored to its environment. The living form would be optimally designed with no poorly working or even nonworking parts. However, evolution suggests that living things are rather like the engineer in the field as they adapt to their environment. When the environment changes, living things must adapt to thrive in new situations. Living things adapt; they do not have the luxury of sitting down to design the body that would be most perfectly attuned with their environment. In some cases, mutations affecting the existing organism may make it better suited to the new environment. If so, the enhanced nature of the organism allows better survival and thus strengthens the likelihood of reproduction. The new features of the organism are passed on to the descendants by natural selection. In other cases, a new feature may appear through a mutation that might not offer an advantage.

A land mammal might find it convenient to return to living in water. As time goes on, mutations may introduce features in the species that make it better adapted to the aquatic life. But its adaptation will be limited by its original ecological place as an air-breathing land mammal. The animal will continue to breathe air, even though the ability to extract oxygen from the water would be a stronger adaptation to a marine existence. According to the analysis of paleontologists, this evolutionary process is evident in many of the descendants of land creatures that have returned to the water: Most of them have continued to be air breathers. Reptiles such as turtles and sea snakes, mammals such as whales and seals, and sea birds all have a remarkable ability to hold their breath and function underwater for extended periods of time. Ultimately, they all come to the surface to breathe. Amphibians, in some cases, have maintained the structure of the original gills of aquatic creatures for part of their evolution. Some salamanders indeed retain gills throughout their lives. However, many salamanders, as well as frogs, become air-breathing adults in spite of spending a lot of time in the water.

Consider the panda, for example. Taxonomists generally agree that pandas are bears or at very least closely related to bears. As with the ancestors of bears, the ancestors of modern pandas were originally omnivorous animals, eating both vegetation and meat. As the result of changes in the environment, pandas eventually specialized and their diet shifted to eating bamboo. However, their digestive system was better attuned to eating meat and not very well suited for a wholly vegetarian diet. Indeed, the genetic composition of modern pandas still provides all the enzymes for carnivorous digestion but not for the cellulose of their bamboo diet.[2] Further, the paws of panda ancestors were not well adapted to coping with a bamboo diet. In time, pandas evolved a special "thumb" that helps them to strip the leaves off bamboo plants. The explanation given by scientists tracking panda evolution is

2. E. S. Dierenfeld et al., "Utilization of Bamboo by the Giant Panda," *Journal of Nutrition* 112 (1982): 636–41.

that they actually lost their thumb in the course of evolution, like bears did. However, in the case of the panda, a nodule of the wrist bone developed into a thumb-like projection to help in the job of stripping the bamboo leaves.

Also, because the diet of the panda's ancestors was once rich in animal proteins, the modern panda has a short gut. This would be fine if it were a meat eater. Even modern pandas sometimes supplement their bamboo diets by scavenging the carcasses of dead animals. Modern pandas have not yet developed an efficient and long intestinal tract that enlists bacteria to help break down the cellulose so that the nutrients from vegetation become usable. Recent studies of panda digestion indicated some use of bacteria in breaking down their diet. However, the relatively short period of time that it takes the food to traverse the digestive tract of pandas (only eight to ten hours from ingestion to excretion) limits the amount of benefit the vegetation can provide.[3] The digestion of pandas is still very inefficient compared to most herbivores, breaking down less than 20 percent of the dry matter equivalent. The result is that pandas have adopted strategies to compensate for the low calorie extraction from their diet. Pandas spend long periods each day eating (up to sixteen hours a day). Further, their physical activity is low and, as a result, they burn fewer calories.

Had the panda been designed to live and operate optimally in its habitat, it would have been given a proper vegetarian digestive tract to more efficiently break down the bamboo leaves it depends upon. If the panda had been designed specifically as a bamboo eater, it would more efficiently extract many more calories from the bamboo it eats each day. Instead, pandas constantly teeter on the edge of starvation. The leaf-stripping "thumb" is in reality a projected wrist bone, not one of the digits on the paw. It does the basic job it needs to do, albeit in an inefficient way. The panda ably illustrates the amazing principles of adaptation in the evolutionary process.

We observe the steady development of life by evolution through natural selection rather than a single creation of all living things at one time. Does this mean that humans are just an accident in the scheme of things? Some argue this, but it is not an inevitable conclusion from the acceptance of evolution.

3. Lifeng Zhua et al., "Evidence of Cellulose Metabolism by the Giant Panda Gut Microbiome," *Proceedings of the National Academy of Sciences* 108.43 (2011): 17714–19.

In his remarkable book *Wonderful Life* Gould examines the exotic creatures discovered in the Burgess Shale.[4] These early Cambrian creatures are not only different from any known modern animals, but in a number of cases they also represent phyla unknown among modern living things. In other words, they cannot be categorized with any modern classes of animals. As an evolutionist, Gould explained this situation as due to a rapid development of living creatures in this period — the so-called "Cambrian explosion." The sudden increase in available and habitable space in this period invited biodiversity to fill these new and relatively empty environmental niches. New species rapidly evolved in a huge variety of different ways where the environment was amenable. However, as these environments proved unsustainable, many of them died out in turn as the environment changed and became hostile. Yet some survived and continued to evolve, resulting in the current classes and species of animals.

Gould argues that our modern world is more or less an accident. That is, if other circumstances had affected the early Cambrian creatures, some of those that died would have survived, while some of those that survived would have died. In the end, the chain of evolution would have led to a different set of modern creatures, which probably would not have included humans. As Gould put it, "The wonder of life is that it need not have happened. Replay the tape of life again, starting with the Burgess Shale, and a different set of survivors, worthy of our science-fiction dreams, would grace our planet today. We would not be among them."[5] Many evolutionists have accepted this picture. However, some have argued strongly against Gould to propose another scenario. I draw particular attention to the Cambridge paleontologist Simon Conway Morris, whose work on the Burgess Shale Gould depended upon.

Conway Morris, in his book *Life's Solution: Inevitable Humans in a Lonely Universe,* argues that the constraints of natural laws must be taken into account in our expectations of what is possible in terms of evolution.[6] Just as the "Goldilocks effect" allows life only within narrow physical limits, so the

4. Located in the Canadian Rockies of British Columbia, this is one of the world's most celebrated fossil fields famous for the exceptional preservation of the soft parts of its fossils.

5. Gould, *Wonderful Life*, see inside the front flap.

6. Simon Conway Morris, *Life's Solution: Inevitable Humans in a Lonely Universe* (Cambridge: Cambridge University Press, 2003).

laws of physics, chemistry, and biology would constrain the development of life within certain parameters. In his view, even if the chain of evolution from the Cambrian explosion to the present day was altered with different environmental factors, the resulting life-forms would not be hugely different from those actually alive today. Of course, they would not be exactly the same. The world would, superficially at least, appear much different from the present life-forms that populate our planet. However, in Conway Morris's analysis, recognizable forms of life would still be evident. There could well still have appeared some sort of intelligent creature with many of the main characteristics seen in today's humans. A divergent evolutionary pathway might have produced corresponding intelligence that would have a different bodily form, but the development of consciousness, intelligence, and even culture would probably still have evolved.

Conway Morris argues his case from a scientific point of view. He is an evolutionary biologist and makes full use of evolutionary theory to construct his interpretation of the scientific data. Thus, he argues – contrary to evolutionary thinkers such as Gould – that humans were "inevitable." Given other environmental factors, these humanlike creatures would undoubtedly look different from the modern humans, but they would possess many of the unique characteristics seen in humans today.

The issue of how living things got here has often been cast as a battle between religion and science. We have seen above that this is a false dichotomy. It is true that some scientists characterize the debate this way, implying that science and religion are incompatible. However, this is a minority view in the scientific community. Many scientists are not interested in discussing religion or broader philosophical issues. They prefer to focus on their areas of expertise and specialized research. Other scientists show interest in the philosophical and even the religious questions and implications of some of their work. For example, in a recent interview the Astronomer Royal and head of the Royal [Scientific] Society, Lord Martin Rees, said that he believed science and religion belonged to different domains and could coexist peacefully. He also noted that theologians who take their theology seriously do not attempt to use theology to explain the mysteries of science. We have seen that there are many scientists who find no contradiction between their scientific work and personal faith, beliefs, or practices. Likewise, many of these scientists also find no contradiction between accepting evolution as a scientifically accurate description of origins and embracing their personal religious beliefs.

The problem is not one of science at all. For some believers, the real issue is how to understand the Bible. As I attempted to demonstrate with my own biography in chapter 1, serious study of the Bible does not demand the abandoning of religious commitments. Accepting the Bible does not mean throwing out the findings of science. It is possible to maintain faith, even though one's understanding of the Bible may need to grow.

In a moving account (originally titled *Evolving in Monkey Town*),[7] Rachel Held Evans relates how she grew up as a fundamentalist in Dayton, Tennessee, where the infamous Scopes trial took place. She graduated from the local Bryan College. Evans's crisis of faith came when she saw an Afghani woman brutally executed on video. Her concern was not primarily Darwinian evolution, which she had firmly rejected up to then. Yet Darwinian evolution serves as a powerful metaphor for her own personal evolution in belief and understanding. Although still a Christian, her perspective on the Bible has changed significantly, as she graphically describes (for one thing, she accepts the scientific basis for evolution).

The Bible, when carefully studied in its original languages and in the context of the literature from the Mediterranean and the ancient Near East, opens new doors of understanding. The literal seven days of creation or the meaning of "kinds" – that creationists and some believers assume are the plain meaning of Scripture – are simply not possible meanings of the biblical text. For me personally, such a realization was a process, and I did not reach these conclusions lightly or without considerable thought and soul-searching.

One problem of those who wish to impose a rather uninspired literal reading on the Bible's creation stories is that this conjures up a situation where God deceives humanity in the way the natural world was created. Fossils and the geological record would be apparently old, so that geologists would be forced to reach false conclusions about the age of the earth and how it developed over time. The past half century of historical study has

7. Rachel Held Evans, *Evolving in Monkey Town: How a Girl Who Knew all the Answers Learned to Ask the Questions* (Grand Rapids: Zondervan, 2010), reprinted as *Faith Unraveled: How a Girl Who Knew all the Answers Learned to Ask the Questions* (Grand Rapids: Zondervan, 2014).

taught us quite a bit more about how the Bible was written and its meaning in its own historical context. To read the Bible faithfully, readers need to take into account the linguistic, literary, historical, and cultural background that is now readily available to careful readers, even if this evidence forces us to drop certain *cherished assumptions*.

As a young man, I assumed that the Bible was inerrant in the autographs, without any error in the original words put to parchment by inspired authors. However, a closer study of the biblical text led me to a more nuanced appreciation of the Bible. There is a history of the transmission of the biblical texts. The text had often been revised, altered, and even suffered errors in copying. Many of the sources of the translations of the Bible today are from manuscripts copied in the medieval period. There are of course much earlier manuscripts found among the Dead Sea Scrolls as well as from early translations of the biblical text into other languages. This history is complex and careful scholars have done a good job piecing together the most ancient form of these biblical texts. However, we do not have anything like the "original autographs."

Assumptions of "inerrancy" are patently untenable when we recognize the history of the text. The argument that the text is inerrant, perfect, and without flaw is possible only when history is ignored. In the case of many books of the Bible, there were never any autographs or originals in the sense imagined by some. The text we have today was composed over time, having material constantly added to it, being periodically edited, revised, and updated. In the few cases where there was a single-authored manuscript, such as perhaps with the book of Qohelet (Ecclesiastes), there are still errors in the text from imperfect copying. The Dead Sea Scrolls manuscripts demonstrate the fluidity of texts.

Worse yet is the view often found among creationists that the Bible is somehow a scientific or historical handbook, meant to preserve us from the human errors of scientific study. The Bible reflects a very human history in its language and literary forms. This does not deny divine inspiration. Historically, few theologians believed that inspiration occurred by some sort of divine dictation or an inerrant writing dropped from heaven. Writers wrote of their encounter with God, but that encounter was presented in the language, memory, thought forms, and understandings of the very human writer in his own time and place, limited by his (perhaps occasionally "her") perspective as a person living in the ancient Near East among a small and often subject people. These writers did not have some unique knowledge that only we today can understand. They did not have scientific knowledge that subsequent generations did not understand until the present when – as if by magic – we suddenly understand what they were telling us. No, they wrote as ancients with language and images that would have made imaginative sense to their contemporaries. They did not have our scientific knowledge; they wrote about what they knew.

Today we all accept the benefits of science: medical advances, computers, smartphones, Internet, transportation systems, and increased agricultural production.

Evolution is not on the fringe of the real science we all profit from; evolution is widely accepted because it is part and parcel of the tapestry of science today, well supported by a number of different scientific branches. Many of these advances we enjoy have depended on the science of evolution. Multiple fields converge in undergirding and furthering the understanding of the evolutionary processes that have taken place: these include embryology, cytology (cell biology), and genetics, not to mention zoology, botany, geology, and paleontology. And multiple numbers of these scientists believe in God and have a religious faith.

Evolutionary biologists, paleontologists, geologists, and scientists in related areas all go about their tasks just as do researchers in other areas of science. Evolution provides a theoretical framework, just as the atomic theory does to physicists and chemists and the big bang does to cosmologists; that is, after more than one hundred and fifty years of work by thousands of paleontologists and related scientists – with countless tests, experiments, and cross-checks – the theory of evolution has stood the test of time. It is constantly being subjected to examination, just as with any other scientific theory. Scientists use evolution as a framework because its explanatory power effectively answers to the evidence. This is exactly how it is with other fields of science. Atoms, molecules, and quantum theory are used because they answer to the evidence, and experiments designed to test if they are real routinely confirm that subatomic theory is an accurate picture of the way things are. It may be that one day a better theory will replace evolution as it now stands. But any new theory of origins that may emerge someday is likely to subsume Darwinian evolution within itself rather than abandon it, in the same way that Newtonian physics has been encompassed in modern physics rather than just being jettisoned.

Of course, there remain many unanswered questions relating to science and evolution or evolution as it relates to the Bible. Most scientists are forthright about the problems, gaps in knowledge, and uncertainties. In the same way, biblical scholars are willing to differentiate between what has a solid basis in the evidence and what is a matter of interpretation or even speculation. The problem, as in many fields of knowledge, is when someone who is not trained in a field makes assumptions for which they have far too little evidence.

Hundreds or even thousands of Christian and Jewish geologists see no conflict between their science of the earth and the Bible. They do not feel the need to reject the findings of geology. They are content to accept that the geological column that has been painstakingly worked out by fellow scientists is a good approximation of the development of life. These scientists of faith are not compelled to make a worldwide flood overlap with the evidence from fossils or geology. As we have seen, even many creationists now accept the validity of the geological column. Sadly, some creationists continue to argue that traditional geology is a falsehood inflicted on an innocent world. Invariably, those who do so today in spite of all the contrary evidence are non-geologists. What that geological record shows is a gradual development in the complexity of life-forms. The origin of life itself is still a mystery. The evidence tells us that for billions of years only microscopic creatures existed. It was only less than a billion years ago that multicelled life-forms appeared in a period referred to as the "Cambrian explosion." The development of life is there in the rocks of the earth for all to see, if only one will stop twisting the biblical narrative into an apologetic that needlessly rejects it.

"Surely," some will say, "all the various flood myths and stories around the world must go back to a universal flood that swallowed the whole earth. The flood of Noah in the Bible is attested in many ancient stories." The literary evidence does not compel this conclusion. First, floods are an almost universal human experience. Most regions have had devastating floods of some sort in historical memory. It would be surprising if these experiences did not get passed down in stories and legends of some sort. Some of these may resemble the story of Noah in the Bible in some respects. Second, the variety of flood legends in the ancient Near East do have a literary relationship that we have traced earlier. Most biblical scholars agree that the story of Noah in Genesis 6–9 was borrowed and adapted from the Mesopotamian version – with emphasis on "adapted." The outline of the story in Genesis is very similar to earlier Mesopotamian accounts, but the writer has completely rewritten it and brought to it Hebrew thought forms and theology to make it his own. The biblical story of Noah, unlike the other related stories, supports the author's monotheistic beliefs.

The different versions are not independent accounts but seem to have developed from an account in Mesopotamia. The story may have been based on a historical incident, though it is difficult to say precisely what historical event gave shape to the story. One popular theory is that some sort of mass flooding occurred in the area now known as the Black Sea, which would have lingered long in the ancestral memories of those who lived in the region. The geological evidence, however, does not support a literal worldwide flood, and this is the conclusion of geologists who are themselves people of faith. Given that the story in the Bible is adapted from ancient legend to tell us something about God's character, most biblical scholars likewise accept that the story is a theological tale rather than a historical one.

People of faith have long struggled with the question of theodicy, or questions about the justice of God. Why do the innocent suffer? This is also an important biblical theme. It is one of the central concerns of the book of Job. Some people of faith will have difficulty accepting that their belief in God is compatible with the general concept of evolution. How could a benevolent, loving creator form life through the ages by means of survival of the fittest? However, this question reflects a misunderstanding of evolution. Often attributed to Darwin, the phrase was in reality coined by others.[8] If we think of organisms as merely in an endless cycle of competition, this is not an accurate picture of how evolution works. An organism may be well suited to one environment but poorly suited to another. Natural selection ensures that the individuals within a species better suited to the environment or situation will usually survive. There is nothing nefarious about this. The pressures of environment have helped some specimens develop characteristics that better enable them to survive

8. Darwin did eventually use it, but as a synonym for "natural selection." See the discussion in Stephen Jay Gould, "Darwin's Untimely Burial," *Natural History* 85.7 (1976): 24–30.

in hostile conditions. This is what makes, over time, so many species so well suited to their environments.

Dawkins has introduced a term that added needless grief for people of faith to accept the findings of evolution, namely the term "selfish gene." This controversial idea, by no means universally accepted by scientists (though some regard it very favorably), pictures natural selection taking place at the genetic level rather than merely at the level of species and adaptability. Some understand this idea to rule out certain spiritual ethical concepts like altruistic behavior. However, a number of those who accept the idea of natural selection at the level of genetics do not accept that this somehow does away with selflessness. Natural selection can well be the mechanism of creation without undermining the call to love your neighbor as yourself.

To accept that God created by using evolution might actually alleviate some of the questions over his goodness. In this way of thinking, painful things may happen in this world because God allows the world to take the course of creation through natural selection. Diseases arise as bacteria and viruses evolve and change. Genetic disorders occur as a result of imperfections in the DNA process.

It is clear from numerous biblical passages that the God of biblical faith is not responsible for human evil. Free will in human terms allows people to make good and bad decisions. Each of us is a free moral agent and is responsible for our choices, as well as for the consequences of our actions. It would be foolish to blame God or others for our own moral failures. It seems that the good gained by learning through our moral lapses, even in suffering, is a greater good in terms of biblical faith than the loss of freedom that would be needed to prevent us from making painful or hurtful choices.[9]

Many evangelicals find this disturbing because it does not fit with their theology and their evangelical faith, particularly as it relates to human beings. At the risk of repetition, I would like to draw attention to one particular recent discussion in *The Evolution of Adam: What the Bible Does and Doesn't Say about Human Origins*. There, Enns examines the relevant

9. See Alister E. McGrath, *Intellectuals Don't Need God & Other Modern Myths: Building Bridges to Faith Through Apologetics* (Grand Rapids: Zondervan, 1993); Oord, *Polkinghorne Reader*; John Hick, *Evil and the God of Love*, 2nd ed. (London: Macmillan, 1977); Laing, *Even Dawkins Has a God*; Colin Tudge, *Why Genes Are Not Selfish and People Are Nice: A Challenge to the Dangerous Ideas That Dominate Our Lives* (Edinburgh: Floris Books, 2013).

passages in Genesis and the letters of Paul on the question of Adam as the first man. Enns is concerned to show what has long been known among theologians: Paul does not necessarily use the statements of Genesis in a way that follows their obvious meaning. For example, Paul does not quote from the Hebrew text but usually from the Greek translation, which sometimes differs in minor and even major ways from the Hebrew. Paul's understanding is clear: "The universal and self-evident problem of death, the universal and self-evident problem of sin, the historical event of the death and resurrection of Christ."[10] Enns goes on to give a thoughtful assessment of Paul's understanding of Adam and how that understanding remains at the bedrock of Christianity, without necessarily subscribing to a literal Adam and Eve, complete with a serpent in the Garden of Eden around six thousand years ago.

The Bible is a book from faith communities for faith communities that teaches about theology and morality while providing an understanding of the relationship between humanity and God. The Bible cannot teach about the history of rocks or science or technology, as the Bible arises out of a context where modern understandings of these areas of knowledge would have had no reason to emerge.

Most people of faith relate to their world with resources beyond their sacred texts. While some may pride themselves on being "faithful to the Bible," it is readily apparent that beliefs, traditions, and practices that try to force a particular sort of literalism on the text do not in fact arise from the text. We rely on our sacred texts to give us spiritual grounding but need other human endeavors and knowledge to grapple with living in the natural world.

It is not always self-evident how the Bible can inform the faith of believers today. We face questions and problems in the present that were never dreamed of when the Bible was written. Through careful study, principles and guidelines can help us read biblical texts as relevant to our contemporary situation. It is clear that trying to shoehorn the ancient worldview of some of the biblical language into a modern "shoe" is not being faithful to the text. Rather, it twists its plausible meaning and makes it say things that are irrelevant to a proper understanding of either the Bible or the modern world.

10. Enns, *The Evolution of Adam*, 137.

In the same way, we do violence to the Bible if we attempt to make it speak in pseudoscientific ways about modern science and technology. The Bible is not a scientific treatise and cannot offer that sort of information about the world. We need to let science be science and let scientists do their job. The means by which God created through evolution is for science to describe as it describes the property of atoms or the nature of gravity. Along with the many Christian theologians noted above, we can safely accept these findings and find ways to articulate theology that allow the language and descriptions of evolution to be true to themselves.

God somehow – in his own way – is the author of creation and all that exists. That is what the Bible teaches. The question of how God creates, however, is a matter for science. We need to let scientists exercise their God-given mind and expertise in discovering the wonders of how this happens. In that freedom, we can find intellectual and spiritual wonder at the creation before us as it has evolved.

BIBLIOGRAPHY

Alexander, Denis. *Beyond Science*. Philadelphia: A. J. Holman, 1972.

———. *Creation or Evolution: Do We Have to Choose?* 2nd ed. Oxford: Monarch Books, 2014.

———. *Rebuilding the Matrix: Science and Faith in the 21st Century*. Grand Rapids: Zondervan, 2003.

Allmon, Warren D. "The 'God Spectrum' and the Uneven Search for a Consistent View of the Natural World." Pages 180–239 in *For the Rock Record: Geologists on Intelligent Design*. Edited by Jill S. Schneiderman and Warren D. Allmon. Berkeley: University of California Press, 2009.

Applegate, Kathryn, and J. B. Stump, eds. *How I Changed My Mind about Evolution: Evangelicals Reflect on Faith and Science*. Downers Grove, IL: InterVarsity Press Academic, 2016.

Asher, Robert J. *Evolution and Belief: Confessions of a Religious Paleontologist*. Cambridge: Cambridge University Press, 2012.

Ayala, Francisco J. *Darwin and Intelligent Design*. Minneapolis: Fortress Press, 2006.

Baab, Karen L., Peter Brown, Dean Falk, Joan T. Richtsmeier, Charles F. Hildebolt, Kirk Smith, and William Jungers. "A Critical Evaluation of the Down Syndrome Diagnosis for LB1, Type Specimen of *Homo floresiensis*." *PLOS One* (2016): 1–32.

Bajpai, Sunil, and J. G. M. Thewissen. "A New, Diminutive Eocene Whale from Kachchh (Gujarat, India) and Its Implications for Locomotor Evolution of Cetaceans." *Current Science* 79.10 (2000): 1478–82.

Bajpai, Sunil, J. G. M. Thewissen, and Ashok Sahni. "The Origin and Early Evolution of Whales: Macroevolution Documented on the Indian Subcontinent." *Journal of Biosciences* 34.5 (2009): 673–86.

Bakker, Robert T. *The Dinosaur Heresies: New Theories Unlocking the Mystery of the Dinosaurs and Their Extinction*. New York: William Morrow, 1986.

Barrett, Matthew, and Ardel B. Caneday, eds. *Four Views on the Historical Adam*. Counterpoints: Bible and Theology. Grand Rapids: Zondervan, 2013.

Berger, Lee R., John Hawks, Darryl J. de Ruiter, Steven E. Churchill, Peter Schmid, Lucas K. Delezene, Tracy L. Kivell, et al. "*Homo naledi*, A New Species of the Genus *Homo* from the Dinaledi Chamber, South Africa." *eLife* (2015).

Bergh, Gerrit D. van den, Yousuke Kaifu, Iwan Kurniawan, Reiko T. Kono, Adam Brumm, Erick Setiyabudi, Fachroel Aziz, and Michael J. Morwood. "*Homo floresiensis*-like Fossils from the Early Middle Pleistocene of Flores." *Nature* 534 (2016): 245–48.

Berry, R. J., ed. *Real Science, Real Faith*. Crowborough: Monarch Publications, 1991.

———. *Real Scientists, Real Faith*. Oxford: Monarch Books, 2009.

Bhart-Anjan, S. Bhullar, Zachary S. Morris, Elizabeth M. Sefton, Atalay Tok, Masayoshi Tokita, Bumjin Namkoong, Jasmin Camacho, David A. Burnham, and Arhat Abzhanov. "A Molecular Mechanism for the Origin of a Key Evolutionary Innovation, the Bird Beak and Palate, Revealed by an Integrative Approach to Major Transitions in Vertebrate History." *Evolution* 69.7 (2015): 1665–77.

Birx, H. James. *Interpreting Evolution: Darwin and Teilhard de Chardin*. Buffalo, NY: Prometheus Books, 1991.

Black, Matthew. *The Book of Enoch or I Enoch: A New English Edition with Commentary and Textual Notes*. Studia in Veteris Testamenti Pseudepigrapha 7. Leiden: Brill, 1985.

Blocher, Henri. *In the Beginning: The Opening Chapters of Genesis*. Translated by David G. Preston. Downers Grove, IL: InterVarsity Press, 1984.

Bowler, Peter J. *Monkey Trials and Gorilla Sermons: Evolution and Christianity from Darwin to Intelligent Design*. New Histories of Science, Technology, and Medicine. Cambridge: Harvard University Press, 2007.

Boyd, Robert, and Joan B. Silk. *How Humans Evolved*. 7th ed. New York: W. W. Norton, 2015.

Brown, William P. *The Seven Pillars of Creation: The Bible, Science, and the Ecology of Wonder*. Oxford: Oxford University Press, 2010.

Burstein, Stanley Mayer. *The* Babyloniaca *of Berossus*. Malibu, CA: Undena Publications, 1978.

Carroll, Robert. *The Rise of Amphibians: 365 Million Years of Evolution*. Baltimore: Johns Hopkins University Press, 2009.

Cavanaugh, David P., Todd Charles Wood, and Kurt P. Wise. "Fossil Equidae: A Monobaraminic, Stratomorphic Series." Pages 143–53 in *Proceedings of the Fifth International Conference on Creationism*. Edited by Robert L. Ivey. Pittsburgh: Creation Science Fellowship, 2003.

Cavanaugh, William T. *The Myth of Religious Violence: Secular Ideology and the Roots of Modern Conflict*. Oxford: Oxford University Press, 2009.

Cavanaugh, William T., and James K. A. Smith, eds. *Evolution and the Fall*. Grand Rapids: Eerdmans, 2017.

Chiappe, Luis M., and Lawrence M. Witmer, eds. *Mesozoic Birds: Above the Heads of Dinosaurs*. Berkeley: University of California Press, 2002.

Churchill, Morgan, Manuel Martinez-Caceres, Christian de Muizon, Jessica Mnieckowski, and Jonathan H. Geisler. "The Origin of High-Frequency Hearing in Whales." *Current Biology* 26.16 (2016): 2144–49.

Clack, Jennifer A. *Gaining Ground: The Origin and Evolution of Tetrapods.* 2nd ed. Bloomington: Indiana University Press, 2012.

Collins, Francis S. *The Language of God: A Scientist Presents Evidence for Belief.* London: Pocket Books, 2007.

Conway Morris, Simon. *The Crucible of Creation: The Burgess Shale and the Rise of Animals.* Oxford University Press, 1998.

———. *Life's Solution: Inevitable Humans in a Lonely Universe.* Cambridge: Cambridge University Press, 2003.

Cross, Frank Moore, Donald W. Parry, Richard J. Saley, and Eugene Ulrich. *Qumran Cave 4. XII: 1–2 Samuel.* DJD 17. Oxford: Clarendon, 2005.

Cunningham, Conor. *Darwin's Pious Idea: Why the Ultra-Darwinists and Creationists Both Get It Wrong.* Grand Rapids: Eerdmans, 2010.

Currie, Philip J., Eva B. Koppelhus, Martin A. Shugar, and Joanna L. Wright, eds. *Feathered Dragons: Studies on the Transition from Dinosaurs to Birds.* Bloomington: Indiana University Press, 2004.

Darwin, Charles. *The Origin of Species by Means of Natural Selection, or The Preservation of Favoured Races in the Struggle for Life.* 6th ed. London: John Murray, 1897.

Davidson, Gregg R. *When Faith and Science Collide: A Biblical Approach to Evaluating Evolution and the Age of the Earth.* Oxford, MS: Malius Press, 2009.

Davies, Paul. *The Goldilocks Enigma: Why Is the Universe Just Right for Life?* London: Allen Lane, 2006.

Dawkins, Richard. *The God Delusion.* London: Bantam, 2006.

Day, John. *God's Conflict with the Dragon and the Sea.* Washington, DC: Catholic University of America Press, 1985.

Dierenfeld, E. S., H. F. Hintz, J. B. Robertson, P. J. Wan Soest, and O. T. Oftedal. "Utilization of Bamboo by the Giant Panda." *Journal of Nutrition* 112 (1982): 636–41.

Dietrich, Manfried, Oswald Loretz, and Joaquín Sanmartín, eds. *Die keilalphabetischen Texte aus Ugarit, Ras Ibn Hani und anderen Orten/The Cuneiform Alphabetic Texts from Ugarit, Ras Ibn Hani and Other Places.* 3rd ed. Münster: Ugarit-Verlag, 2013.

Dirks, Paul H. G. M., Lee R. Berger, Eric M. Roberts, Jan D. Kramers, John Hawks, Patrick S. Randolph-Quinney, Marina Elliott, et al. "Geological and Taphonomic Context for the New Hominin Species *Homo naledi* from the Dinaledi Chamber, South Africa." *eLife* (2015).

Dowd, Michael. *Thank God for Evolution: How the Marriage of Science and Religion Will Transform Your Life and Our World.* London: Penguin, 2009.

Drouin, Guy, Jean-Rémi Godin, and Benoît Pagé. "The Genetics of Vitamin C Loss in Vertebrates." *Current Genomics* 12.5 (2011): 371–78.

Dyke, Gareth, and Gary Kaiser, eds. *Living Dinosaurs: The Evolutionary History of Modern Birds*. Chichester: Wiley-Blackwell, 2011.

Eldredge, Niles. *The Triumph of Evolution and the Failure of Creationism*. New York: W. H. Freeman, 2000.

Ellis, Richard. *Aquagenesis: The Origin and Evolution of Life in the Sea*. New York: Viking, 2001.

Enns, Peter. *The Evolution of Adam: What the Bible Does and Doesn't Say about Human Origins*. Grand Rapids: Brazos, 2012.

Evans, Rachel Held. *Evolving in Monkey Town: How a Girl Who Knew All the Answers Learned to Ask the Questions*. Grand Rapids: Zondervan, 2010.

———. *Faith Unraveled: How a Girl Who Knew all the Answers Learned to Ask the Questions*. Grand Rapids: Zondervan, 2014.

Falk, Darrel R. *Coming to Peace with Science: Bridging the Worlds between Faith and Biology*. Downers Grove, IL: InterVarsity Press, 2004.

Finkel, Irving. *The Ark before Noah: Decoding the Story of the Flood*. London: Hodder & Stoughton, 2014.

Fleagle, John G. *Primate Adaptation and Evolution*. 3rd ed. San Diego: Academic Press, 2013.

Flower, William Henry. "On Whales, Past and Present, and Their Probable Origin." Pages 209–31 in *Essays on Museums and Other Subjects Connected with Natural History*. London: Macmillan, 1898.

Foster, Benjamin R. *Before the Muses: An Anthology of Akkadian Literature*. 2 vols. Bethesda, MD: CDL Press, 1993.

Fowler, Thomas B., and Daniel Kuebler. *The Evolution Controversy: A Survey of Competing Theories*. Grand Rapids: BakerAcademic, 2007.

Freedman, Harry, and Maurice Simon, trans. and eds. *The Midrash Rabbah: Genesis*. 3rd ed. 2 vols. London: Soncino Press, 1977.

Fu, Qiaomei, Mateja Hajdinjak, Oana Teodora Molodvan, Silviu Constantin, Swapan Mallick, Pontus Skoglund, Nick Patterson, et al. "An Early Modern Human from Romania with a Recent Neanderthal Ancestor." *Nature* 524 (2015): 216–19.

George, Andrew R. *The Epic of Gilgamesh: The Babylonian Epic Poem and Other Texts in Akkadian and Sumerian*. London: Penguin, 1999.

Gibson, John C. L., and G. R. Driver. *Canaanite Myths and Legends*. 2nd ed. Edinburgh: T&T Clark, 1978.

Gilbert, Scott F., and Ziony Zevit. "Congenital Human Baculum Deficiency: The Generative Bone of Genesis 2:21–23." *American Journal of Medical Genetics* 101.3 (2001): 284–85.

Gingerich, Philip D., Munir ul-Haq, Wighart von Koenigswald, William J. Sanders, B. Holly Smith, and Iyad S. Zalmout. "New Protocetid Whale from the Middle Eocene of Pakistan: Birth on Land, Precocial Development, and Sexual Dimorphism." *PloS One* 4.2 (2009): 1–20.

Gingerich, Philip D., B. Holly Smith, and Elwyn L. Simons. "Hind Limbs of Eocene *Basilosaurus*: Evidence of Feet in Whales." *Science* 249 (1990): 154–57.
Gingerich, Philip D., S. Mahmood Raza, Muhammad Arif, Mohammad Anwar, and Xiaoyuan Zhou. "Partial Skeletons of *Indocetus Ramani* (Mammalia, Cetacea) from the Lower Middle Eocene Domanda Shale in the Sulaiman Range of Punjab (Pakistan)." *Contributions from the Museum of Paleontology, The University of Michigan* 28.16 (1993): 393–416.
Gingerich, Philip D., S. Mahmood Raza, Muhammad Arif, Mohammad Anwar, and Xiaoyuan Zhou. "New Whale from the Eocene of Pakistan and the Origin of Cetacean Swimming." *Nature* 368 (1994): 844–47.
Gogel, Sandra Landis. *A Grammar of Epigraphic Hebrew*. SBLSBS 23. Atlanta: Scholars Press, 1998.
Gómez Aranda, Mariano. "Aspectos científicos en el Comentario de Abraham ibn Ezra al libro de Job." *Henoch* 23 (2001): 81–96.
———. *El Commentario de Abraham ibn Ezra al Libro de Job: Edición crítica, traducción y estudio introductorio*. Consejo superior de investigaciones científicas instituto de filología, Serie A: Literatura Hispano-Hebrea 6. Madrid: Consejo superior de investigaciones científicas instituto de filología, 2004.
Gould, Stephen Jay. "Darwin's Untimely Burial." *Natural History* 85.7 (1976): 24–30.
———. *Rock of Ages: Science and Religion in the Fullness of Life*. Library of Contemporary Thought. New York: Ballantine Publishing, 1999.
———. *Wonderful Life: The Burgess Shale and the Nature of History*. London: Hutchinson Radius, 1989.
Grabbe, Lester L. *Ancient Israel: What Do We Know and How Do We Know It?* London: T&T Clark International, 2007.
———. "Elephantine and the Torah." Pages 125–35 in *In the Shadow of Bezalel: Aramaic, Biblical, and Ancient Near Eastern Studies in Honor of Bezalel Porten*. Edited by Alejandro F. Botta. CHANE 60. Leiden: Brill, 2013.
———. *Etymology in Early Jewish Interpretation: The Hebrew Names in Philo*. BJS 115. Atlanta: Scholars Press, 1988.
———. "Mighty Oaks from (Genetically Manipulated?) Acorns Grow: The Chronicle of the Kings of Judah as a Source of the Deuteronomistic History." Pages 154–73 in *Reflection and Refraction: Studies in Biblical Historiography in Honour of A. Graeme Auld*. Edited by Robert Rezetko, Timothy Lim, and W. Brian Aucker. VTSup 113. Leiden: Brill, 2006.
Graffin, Gregory W., and William B. Provine. "Evolution, Religion and Free Will." *American Scientist* 95.4 (2007): 294–97.
Hallo, William W., ed. *The Context of Scripture*, 3 vols. Leiden: Brill, 1997–2002.
Harmand, Sonia, Jason E. Lewis, Craig S. Feibel, Christopher J. Lepre, Sandrine Prat, Arnaud Lenoble, Xavier Boës, et al. "3.3-Million-Year-Old Stone Tools from Lomekwi 3, West Turkana, Kenya." *Nature* 521 (2015): 310–15.
Harris, Matthew P., Sean M. Hasso, Mark W. J. Ferguson, and John F. Fallon. "The

Development of Archosaurian First-Generation Teeth in a Chicken Mutant." *Current Biology* 16 (2006): 371–77.

Harris, William V. *Ancient Literacy*. Cambridge: Harvard University Press, 1989.

Haught, John F. *God after Darwin: A Theology of Evolution*. Boulder, CO: Westview Press, 2000.

Henze, Matthias, ed. *A Companion to Biblical Interpretation in Early Judaism*. Grand Rapids: Eerdmans, 2012.

Hezser, Catherine. *Jewish Literacy in Roman Palestine*. TSAJ 81. Tübingen: Mohr Siebeck, 2001.

Hick, John. *Evil and the God of Love*. 2nd ed. London: Macmillan, 1977.

Hobolth, Asger, Ole F. Christensen, Thomas Mailund, and Mikkel H. Schierup. "Genomic Relationships and Speciation Times of Human, Chimpanzee, and Gorilla Inferred from a Coalescent Hidden Markov Model." *PLoS Genetics* 3.2 (2007): 294–304.

Horowitz, Wayne. *Mesopotamian Cosmic Geography*. MC 8. Winona Lake, IN: Eisenbrauns, 1998.

Hubbard, Troy D., Iain A. Murray, William H. Bisson, Alexis P. Sullivan, Aswathy Sebastian, George H. Perry, Nina G. Jablonski, and Gary H. Perdew. "Divergent Ah Receptor Ligand Selectivity during Hominin Evolution." *Molecular Biology and Evolution* 33.8 (2016).

Jacobsen, Thorkild. *The Sumerian King List*. AS 11. Chicago: University of Chicago Press, 1939.

Jáuregui, Pablo. "Peter Higgs: 'No soy creyente, pero la ciencia y la religión pueden ser compatibles.'" *El Mundo*, December 27, 2012.

Jeeves, Malcolm A., and R. J. Berry. *Science, Life, and Christian Belief: A Survey of Contemporary Issues*. Grand Rapids: Baker Books, 1998.

Jha, Alok. "Peter Higgs Criticises Richard Dawkins over Anti-religious 'Fundamentalism.'" *The Guardian*, December 26, 2012.

Kelley, Patricia H. "Teaching Evolution during the Week and Bible Study on Sunday." Pages 163–79 in *For the Rock Record: Geologists on Intelligent Design*. Edited by Jill S. Schneiderman and Warren D. Allmon. Berkeley: University of California Press, 2009.

Kilmer, Anne D. "The Mesopotamian Concept of Overpopulation and Its Solution as Reflected in Its Mythology." *Orientalia* 41 (1972): 160–77.

Kirk, Geoffrey, John E. Raven, and Malcolm Schofield, eds. *The Presocratic Philosophers: A Critical History with a Selection of Texts*. 2nd ed. Cambridge: Cambridge University Press, 1984.

Kirkpatrick, Patricia G. *The Old Testament and Folklore Study*. JSOTSup 62. Sheffield: JSOT Press, 1988.

Klein, Richard G. *The Human Career: Human Biological and Cultural Origins*. 3rd ed. Chicago: University of Chicago Press, 2009.

Knauf, E. Axel. "Deborah's Language: Judges Ch. 5 in Its Hebrew and Semitic

Context." Pages 167–82 in *Studia Semitica et Semitohamitica*. Edited by Helen Younansardaroud, Josef Tropper, and Bogdan Burtea. AOAT 317. Münster: Ugarit-Verlag, 2005.

Laing, Neil. *Even Dawkins Has a God: Probing and Exposing the Weaknesses in Richard Dawkins' Arguments in "The God Delusion."* Bloomington, IN: WestBow Press, 2014.

Lam, Monica. "Profile/Eugenie Scott/Berkeley Scientist Leads Fight to Stop Teaching of Creationism." *San Francisco Chronicle*, February 7, 2003.

Lambert, Wilfred G., and Alan R. Millard. *Atra-Ḫasīs: The Babylonian Story of the Flood*. Oxford: Clarendon, 1969.

Lamoureux, Denis O. *Evolutionary Creation: A Christian Approach to Evolution*. Cambridge: Lutterworth, 2008.

———. "No Historical Adam: Evolutionary Creation View." Pages 37–88 in *Four Views on the Historical Adam*. Edited by Matthew Barrett and Ardel B. Caneday. Counterpoints: Bible and Theology. Grand Rapids: Zondervan, 2013.

Larson, Edward J., and Larry Witham. "Leading Scientists Still Reject God." *Nature* 394 (1998): 313.

———. "Scientists Are Still Keeping the Faith." *Nature* 386 (1997): 435–36.

Lilley, Christopher, and Daniel Pedersen, eds. *Human Origins and the Image of God: Essays in Honor of J. Wentzel van Huyssteen*. Grand Rapids: Eerdmans, 2017.

Lord, Albert B. *The Singer of Tales*. Cambridge: Harvard University Press, 1960.

McGrath, Alister E. *Intellectuals Don't Need God & Other Modern Myths: Building Bridges to Faith Through Apologetics*. Grand Rapids: Zondervan, 1993.

———. *A Scientific Theology*. 3 vols. London: T&T Clark, 2001–2003.

McGrath, Alister E., and Joanna Collicutt McGrath. *The Dawkins Delusion? Atheist Fundamentalism and the Denial of the Divine*. Downers Grove, IL: InterVarsity Press, 2010.

McKnight, Scot, and Dennis Venema. *Adam and the Genome: Reading Scripture after Genetic Science*. Grand Rapids: Brazos, 2017.

Madar, Sandra I. "The Postcranial Skeleton of Early Eocene Pakicetid Cetaceans." *Journal of Paleontology* 81.1 (2007): 176–200.

Miller, Kenneth R. *Finding Darwin's God: A Scientist's Search for Common Ground between God and Evolution*. New York: Harper Perennial, 1999.

Moreland, J. P., and John Mark Reynolds, eds. *Three Views on Creation and Evolution*. Grand Rapids: Zondervan, 1999.

Morris, Thomas V., ed. *God and the Philosophers: The Reconciliation of Faith and Reason*. Oxford: Oxford University Press, 1994.

Morton, Glenn R. "The Transformation of a Young-Earth Creationist." *Perspectives on Science and Christian Faith* 52 (2000): 81–83.

Naish, Darren. "Fossils Explained 46: Ancient Toothed Whales." *Geology Today* 20.2 (2004): 72–77.

Neusner, Jacob, ed. *The Babylonian Talmud: Translation and Commentary*. Peabody, MA: Hendrickson, 2006.

Ni, Xijun, Daniel L. Gebo, Marian Dagosto, Jin Meng, Paul Tafforeau, John J. Flynn, and K. Christopher Beard. "The Oldest Known Primate Skeleton and Early Haplorhine Evolution." *Nature* 498 (2013): 60–64.

Nickelsburg, George W. E. *Jewish Literature between the Bible and the Mishnah*. 2nd ed. Minneapolis: Fortress, 2005.

Nickelsburg, George W. E., and James C. VanderKam. *1 Enoch 2: A Commentary on the Book of 1 Enoch, Chapters 37–82*. Hermeneia. Minneapolis: Fortress, 2012.

Numbers, Ronald L. *The Creationists: From Scientific Creationism to Intelligent Design*. Rev. ed. Cambridge: Harvard University Press, 2006.

Nummela, Sirpa, S. T. Hussain, and J. G. M. Thewissen. "Cranial Anatomy of Pakicetidae (Cetacea, Mamallia)." *Journal of Vertebrate Paleontology* 26 (2006): 746–59.

Nummela, Sirpa, J. G. M. Thewissen, Sunil Bajpai, Taseer Hussain, and Kishor Kumar. "Sound Transmission in Archaic and Modern Whales: Anatomical Adaptations for Underwater Hearing." *The Anatomical Record* 290 (2007): 716–33.

Oord, Thomas J., ed. *Polkinghorne Reader: Science, Faith and the Search for Meaning*. London: SPCK, 2010.

Pääbo, Svante. "The Contribution of Ancient Hominin Genomes from Siberia to Our Understanding of Human Evolution." *Herald of the Russian Academy of Sciences* 65 (2015): 392–96.

Polkinghorne, John. *From Physicist to Priest: An Autobiography*. London: SPCK, 2007.

Prothero, Donald R. *Evolution: What the Fossils Say and Why It Matters*. New York: Columbia University Press, 2007.

Prüfer, Kay, Fernando Racimo, Nick Patterson, Flora Jay, Sriram Sankararaman, Susanna Sawyer, Anja Heinze, et al. "The Complete Genome Sequence of a Neanderthal from the Altai Mountains." *Nature* 505 (January 2014): 43–49.

Rasmussen, Morten, Xiaosen Guo, Yong Wang, Kirk E. Lohmueller, Simon Rasmussen, Anders Albrechtsen, Line Skotte, et al. "An Aboriginal Australian Genome Reveals Separate Human Dispersals into Asia." *Science* 334 (2011): 94–98.

Reich, David, Richard E. Green, Martin Kircher, Johannes Krause, Nick Patterson, Eric Y. Durand, Bence Viola, et al. "Genetic History of an Archaic Hominin Group from Denisova Cave in Siberia." *Nature* 468 (2010): 1053–60.

Reich, David, Nick Patterson, Martin Kircher, Frederick Delfin, Madhusudan R. Nandineni, Irina Pugach, Albert Min-Shan Ko, et al. "Denisova Admixture and the First Modern Human Dispersals into Southeast Asia and Oceania." *The American Journal of Human Genetics* 89 (2011): 516–28.

Robertson, David A. *Linguistic Evidence in Dating Early Hebrew Poetry*. SBLDS 3. Atlanta: Scholars Press, 1972.

Rollston, Christopher A. *Writing and Literacy in the World of Ancient Israel: Epi-*

graphic Evidence from the Iron Age. ABS 11. Atlanta: Society of Biblical Literature, 2010.

Roughgarden, Joan. *Evolution and Christian Faith: Reflections of an Evolutionary Biologist*. Washington, DC: Island Press, 2006.

Ruse, Michael. *Can a Darwinian Be a Christian? The Relationship between Science and Religion*. Cambridge: Cambridge University Press, 2000.

———. *The Evolution-Creation Struggle*. Cambridge: Harvard University Press, 2005.

Sankararaman, Sriram, Swapan Mallick, Michael Dannemann, Kay Prüfer, Janet Kelso, Svante Pääbo, Nick Patterson, and David Reich. "The Genomic Landscape of Neanderthal Ancestry in Present-Day Humans." *Nature* 507 (2014): 354–57.

Schneider, John R. "Recent Genetic Science and Christian Theology on Human Origins: An 'Aesthetic Supralapsarianism.'" *Perspectives on Science and Christian Faith* 62.3 (2010): 196–212.

Secord, James A., ed. *Charles Darwin, Evolutionary Writings*. Oxford: Oxford University Press, 2008.

Senter, Phil. "Using Creation Science to Demonstrate Evolution 1: Application of a Creationist Method for Visualizing Gaps in the Fossil Record to a Phylogenetic Study of Coelurosaurian Dinosaurs." *Journal of Evolutionary Biology* 23.8 (2010): 1732–43.

———. "Using Creation Science to Demonstrate Evolution 2: Morphological Continuity within Dinosauria." *Journal of Evolutionary Biology* 24.10 (2011): 2197–216.

———. "Vestigial Structures Exist Even within the Creationist Paradigm." *Reports of the National Center for Science Education* 30.4 (2010).

Shubin, Neil H. *Your Inner Fish: The Amazing Discovery of Our 375-Million-Year-Old Ancestor*. New York: Pantheon Books, 2008.

Simpson, George Gaylord. *The Meaning of Evolution*. Rev. ed. New Haven: Yale University Press, 1967.

Sire, Jean-Yves, Sidney C. Delgado, and Marc Girondot. "Hen's Teeth with Enamel Cap: From Dream to Impossibility." *BMC Evolutionary Biology* 8.246 (2008).

Stec, David M. *The Text of the Targum of Job: An Introduction and Critical Edition*. AGJU 20. Leiden: Brill, 1994.

Stringer, Chris. *The Origin of Our Species*. London: Penguin Books, 2011.

Stringer, Chris, and Peter Andrews. *The Complete World of Human Evolution*. London: Thames and Hudson, 2005.

Sutera, Raymond. "The Origin of Whales and the Power of Independent Evidence." *Reports of the National Center for Science Education* 20.5 (2000): 33–41.

Sutikna, Thomas, Matthew W. Tocheri, Michael J. Morwood, E. Wahyu Saptomo, Jatmiko, Rokus Due Awe, Sri Wasisto, et al. "Revised Stratigraphy and Chronology for *Homo floresiensis* at Liang Bua in Indonesia." *Nature* 532 (2016): 366–69.

Thewissen, J. G. M. *The Walking Whales: From Land to Water in Eight Million Years*. Oakland: University of California Press, 2015.

Thewissen, J. G. M., and Sunil Bajpai. "New Skeletal Material of *Andrewsiphius* and *Kutchicetus*, Two Eocene Cetaceans from India." *Journal of Paleontology* 83.5 (2009): 635–63.

———. "Whale Origins as a Poster Child for Macroevolution." *BioScience* 51.12 (2001): 1037–49.

Thewissen, J. G. M., and E. M. Williams. "The Early Radiations of Cetacea (Mammalia): Evolutionary Pattern and Developmental Correlations." *Annual Review of Ecology and Systematics* 33 (2002): 73–90.

Tudge, Colin. *Why Genes Are Not Selfish and People Are Nice: A Challenge to the Dangerous Ideas that Dominate Our Lives*. Edinburgh: Floris Books, 2013.

VanderKam, James C. *The Dead Sea Scrolls Today*. 2nd ed. Grand Rapids: Eerdmans, 2010.

Velikovsky, Immanuel. *Ages in Chaos*. Garden City, NY: Doubleday, 1952.

Venema, Dennis R. "Genesis and the Genome: Genomics Evidence for Human-Ape Common Ancestry and Ancestral Hominid Population Sizes." *Perspectives on Science and Christian Faith* 62.3 (2010): 166–78.

Vernot, Benjamin, Serena Tucci, Janet Kelso, Joshua G. Schraiber, Aaron B. Wolf, Rachel M. Gittelman, Michael Dannemann, et al. "Excavating Neandertal and Denisovan DNA from the Genomes of Melanesian Individuals." *Science* (2016).

Walton, John H. *Genesis 1 as Ancient Cosmology*. Winona Lake, IN: Eisenbrauns, 2011.

———. *The Lost World of Adam and Eve: Genesis 2–3 and the Human Origins Debate*. Downers Grove: InterVarsity Press Academic, 2015.

———. *The Lost World of Genesis One: Ancient Cosmology and the Origins Debate*. Downers Grove: InterVarsity Press Academic, 2009.

Weston, Eleanor M., and Adrian M. Lister. "Insular Dwarfism in Hippos and a Model for Brain Size Reduction in *Homo floresiensis*." *Nature* 459 (2009): 85–88.

Whitcombe, John C., and Henry M. Morris. *The Genesis Flood: The Biblical Record and Its Scientific Implications*. Phillipsburg, NJ: Presbyterian and Reformed Publishing, 1961.

Wiens, Roger C. "Radiometric Dating: A Christian Perspective," www.asa3.org/ASA/resources/Wiens.html.

Wise, Kurt P. "The Origin of Life's Major Groups." Pages 211–34 in *The Creation Hypothesis: Scientific Evidence for an Intelligent Designer*. Edited by J. P. Moreland. Downers Grove, IL: InterVarsity Press, 1994.

———. "Towards a Creationist Understanding of 'Transitional Forms.'" *CEN Technical Journal* 9.2 (1995): 216–22.

Wolgemuth, Ken, Gregory S. Bennett, and Gregg Davidson. "Theologians Need to Hear from Christian Geologists about Noah's Flood." Paper presented at the

Annual Meeting of the Evangelical Theological Society, New Orleans, November 2009.

Wood, Todd Charles. "The Current Status of Baraminology." *Creation Research Society Quarterly* 43 (2006): 149–58.

Wood, Todd Charles, Kurt P. Wise, Roger Sanders, and N. Doran. "A Refined Baramin Concept." *Occasional Papers of the Baraminology Study Group* 3 (2003): 1–14.

Wright, N. T., "Excursus on Paul's Use of Adam." Pages 170–80 in *The Lost World of Adam and Eve: Genesis 2–3 and the Human Origins Debate*. Edited by John H. Walton. Downers Grove, IL: InterVarsity Press Academic, 2015.

Young, Davis A. *The Biblical Flood: A Case Study of the Church's Response to Extrabiblical Evidence*. Grand Rapids: Eerdmans, 1995.

Young, Davis A., and Ralph F. Stearley. *The Bible, Rocks, and Time: Geological Evidence for the Age of the Earth*. Downers Grove, IL: InterVarsity Press, 2008.

Zevit, Ziony. "Was Eve Made from Adam's Rib – or His Baculum?" *BAR* 41.5 (2015): 32–35.

Zhua, Lifeng, Qi Wua, Jiayin Daia, Shanning Zhangb, and Fuwen Weia. "Evidence of Cellulose Metabolism by the Giant Panda Gut Microbiome." *Proceedings of the National Academy of Sciences* 108.43 (2011): 17714–19.

Zimmer, Carl. *At the Water's Edge: Macroevolution and the Transformation of Life*. New York: The Free Press, 1998.

INDEX OF AUTHORS

INDEX OF SUBJECTS

INDEX OF SCRIPTURES